REPORT

Methodology for Constructing a Modernization Roadmap for Air Force Automatic Test Systems

Lionel A. Galway, Rachel Rue, James M. Masters, Ben D. Van Roo, Manuel Carrillo, Amy L. Maletic, John G. Drew

Prepared for the United States Air Force

Approved for public release; distribution unlimited

PROJECT AIR FORCE

The research described in this report was sponsored by the United States Air Force under Contract FA7014-06-C-0001. Further information may be obtained from the Strategic Planning Division, Directorate of Plans, Hq USAF.

Library of Congress Cataloging-in-Publication Data

Methodology for constructing a modernization roadmap for Air Force automatic test systems / Lionel A. Galway ... [et al.].
    p. cm.
  Includes bibliographical references.
  ISBN 978-0-8330-5899-7 (pbk. : alk. paper)
  1. United States. Air Force—Weapons systems—Testing. I. Galway, Lionel A., 1950-

UG633.M3445 2012
358.4'18—dc23

2012004616

The RAND Corporation is a nonprofit institution that helps improve policy and decisionmaking through research and analysis. RAND's publications do not necessarily reflect the opinions of its research clients and sponsors.

RAND® is a registered trademark.

Published 2012 by the RAND Corporation
1776 Main Street, P.O. Box 2138, Santa Monica, CA 90407-2138
1200 South Hayes Street, Arlington, VA 22202-5050
4570 Fifth Avenue, Suite 600, Pittsburgh, PA 15213-2665
RAND URL: http://www.rand.org/
To order RAND documents or to obtain additional information, contact
Distribution Services: Telephone: (310) 451-7002;
Fax: (310) 451-6915; Email: order@rand.org

Preface

The Air Force uses an extensive set of logistics assets and facilities to support training, deployment, employment, and redeployment of air, space, and cyber forces. Recognizing the importance of test equipment to the logistics enterprise, the Air Force would like to develop a roadmap to guide the transition of workloads from legacy automatic test systems (ATSs) to Department of Defense–approved families of ATS equipment. In response, RAND Project AIR FORCE developed a comprehensive enterprise approach for evaluating options for migrating individual weapon system and functional ATS capabilities to a common ATS family.

This report specifically presents our methodology for conducting an economic analysis for moving legacy and future systems to a common family, then illustrates this approach by developing a portion of the enterprise ATS roadmap for some key functional capabilities. The Air Force can further apply this methodology as it develops a comprehensive ATS roadmap.

The report documents the methods and findings of the fiscal year 2009 study, "Air Force Roadmap for Automatic Test Systems." The research reported here was sponsored by Maj Gen Polly A. Peyer, then commander of Warner Robins Air Logistics Center (ALC), and Maj Gen Robert H. McMahon, then AF/A4L, and was conducted within the Resource Management Program of RAND Project AIR FORCE. This report is intended to help inform planners, the test equipment System Program Office (SPO), the acquisition community, the logistics community, and anyone involved with depot level repair or intermediate level repair of components, including those in the Air Reserve Component.

This research is part of a broader research portfolio that addresses improving the Air Force's ability to respond to a dynamic environment. Related publications include

- *A Repair Network Concept for Air Force Maintenance: Conclusions from Analysis of C-130, F-16, and KC-135 Fleets,* by Robert S. Tripp, Ronald G. McGarvey, Ben D. Van Roo, James M. Masters, and Jerry M. Sollinger (MG-919-AF). This monograph describes an analysis of repair network options to support three series of aircraft. It assesses the effect of consolidating certain maintenance tasks at centralized repair facilities and discusses maintenance concepts that integrate wing- and depot-level maintenance processes.
- *Supporting Air and Space Expeditionary Forces: An Expanded Operational Architecture for Combat Support Planning and Execution Control,* by Patrick Mills, Ken Evers, Donna Kinlin, and Robert S. Tripp (MG-316-AF). This monograph expands and provides more detail on several organizational nodes in our earlier work that outlined concepts for an operational architecture for guiding the development of Air Force combat support execution planning and control needed to enable rapid deployment and employment of air and

space expeditionary forces. These are sometimes referred to as combat support execution planning and control processes.

- *Supporting Air and Space Expeditionary Forces: Lessons from Operation Iraqi Freedom*, by Kristin F. Lynch, John G. Drew, Robert S. Tripp, and Charles Robert Roll, Jr. (MG-193-AF). This monograph describes the expeditionary agile combat support experiences during the war in Iraq and compares these experiences with those associated with Joint Task Force Noble Anvil, in Serbia, and Operation Enduring Freedom, in Afghanistan. This report analyzes how combat support performed and how agile combat support concepts were implemented in Iraq, compares current experiences to determine similarities and unique practices, and indicates how well the agile combat support framework performed during these contingency operations.

- *Supporting Air and Space Expeditionary Forces: Analysis of Maintenance Forward Support Location Operations*, by Amanda B. Geller, David George, Robert S. Tripp, Mahyar A. Amouzegar, and Charles Robert Roll, Jr. (MG-151-AF). This monograph discusses the conceptual development and recent implementation of maintenance forward support locations (also known as centralized repair facilities) for the U.S. Air Force.

- *Reconfiguring Footprint to Speed Expeditionary Aerospace Forces Deployment*, by Lionel A. Galway, Mahyar A. Amouzegar, Richard Hillestad, and Don Snyder (MR-1625-AF). This report develops an analysis framework—as a footprint configuration—to assist in devising and evaluating strategies for footprint reduction. The authors attempt to define *footprint* and to establish a way to monitor its reduction.

- *A Combat Support Command and Control Architecture for Supporting the Expeditionary Aerospace Force*, by James Leftwich, Robert S. Tripp, Amanda B. Geller, Patrick Mills, Tom LaTourrette, Charles Robert Roll, Jr., Cauley Von Hoffman, and David Johansen (MR-1536-AF). This report outlines the framework for evaluating options for combat support execution planning and control. The analysis describes the combat support command and control operational architecture as it is now and as it should be in the future. It also describes the changes that must take place to achieve that future state.

RAND Project AIR FORCE

RAND Project AIR FORCE (PAF), a division of the RAND Corporation, is the U.S. Air Force's federally funded research and development center for studies and analyses. PAF provides the Air Force with independent analyses of policy alternatives affecting the development, employment, combat readiness, and support of current and future air, space, and cyber forces. Research is conducted in four programs: Force Modernization and Employment; Manpower, Personnel, and Training; Resource Management; and Strategy and Doctrine.

Additional information about PAF is available on our website:
http://www.rand.org/paf/

Contents

Figures

Tables

Summary

Virtually all the electronics in an Air Force weapon system are tested using automatic test equipment, much of which is unique to that weapon system. However, the Air Force's ATSs are currently beset by increasing hardware and software obsolescence, which is compounded by the number and variety of legacy ATS types. In response to an overall Department of Defense policy, the Air Force is planning to modernize its component repair capabilities, rehosting them on a much smaller number of modern common testing systems. This report focuses on the economic aspect of the rehosting decision, i.e., which component repairs should be rehosted to use resources most efficiently while maintaining repair capabilities.[1]

Our approach was to formulate rehosting decisions for each legacy ATS for each associated unit under test (UUT).[2] These decisions needed to take into account three sets of information: the projected UUT workload for old and new ATSs; the cost to operate, maintain, and sustain the new and legacy systems; and the nonrecurring cost of rehosting the UUTs on a new ATS. To select an optimal rehosting strategy given the costs and constraints, we formulated a mixed integer linear program. As a test case, we selected a set of six avionics ATS types that collectively repair a selected set of 470 UUTs from the B-1B, focusing on a single weapon system with a small number of operating locations to facilitate data collection.

Our analysis showed that the major driver of rehosting cost is that of rewriting the software programs to run on the new ATS to test each UUT (currently estimated at $300,000 to $1 million, depending on the unit's complexity). This implies that good candidates for rehosting should have

1. very high and increasing maintenance costs and obsolescence issues
2. a relatively small number of UUTs repaired.

An example in our analysis is the contrast between the Radar Electronic Warfare (REW) test station and the Depot Automated Test System for Avionics (DATSA). Both have very high sustainment costs (more than $2 million per year), but the DATSA repairs 350 UUTs, while the REW repairs 22.[3] Total estimated rehosting costs would be $22 million for the REW but $111 million for the DATSA. The REW is therefore close to the point at which it

---

[1] Modern common test stations offer other advantages, such as a more flexible workforce and the ability to share workload across depots, which we did not include in our analysis.

[2] The UUT is the electronic system component an ATS tests.

[3] Note that the B-1B avionics suite includes more UUTs than we examined and that the ATSs for some of these, notably the DATSA, repair more UUTs than those in our study.

would be cost-effective to rehost its entire workload over a ten-year period. On the other hand, sustainment costs for the DATSA would have to increase substantially to warrant complete rehosting of its workload.

In some situations, an incremental rehosting strategy may be justified:

- Rehosting a small number of UUTs may substantially reduce the total annual workload for a particular ATS type (the idled systems may be used as spares or a source of compatible components).
- The need for significant software modifications to support sustainment engineering for a legacy ATS may provide an opportunity for transition. In such cases, the software costs are essentially sunk, and rehosting the UUTs on a modern ATS may well be life-cycle cost-effective.

In some cases, numbers of a particular ATS type may be very limited, and those may not be sustainable (catastrophic failure). Rehosting the entire workload may be the only option (although doing so may take some time because of manpower and other constraints).

If the Air Force is to manage ATSs centrally, it needs ongoing access to much better data. For example, there is substantial uncertainty about the costs for legacy ATSs and other data required to use our approach. This is particularly true of software translation costs, for which more detailed information on actual translation costs might help make estimates more accurate.

The long-term benefits of modernized and common testing equipment make a strong case for making common families be the foundation of ATS acquisition on future platforms. The methodology can also be used to calculate a roadmap for rehosting the workloads for current platforms that will be phased out or reduced in the near term (10–20 years).

Acknowledgments

Our sponsors, Maj Gen Polly A. Peyer (WR-ALC/CC) and Maj Gen Robert H. McMahon (AF/A4L), were instrumental in helping us formulate the project and making sure that it came into the PAF portfolio. Our point of contact was Col David French (742 CBSG/CC),[1] the ATS Product Group Manager who has responsibility for moving the Air Force to common families of modernized testers. He and his deputy, Wendy Johnston, provided the overall context for our roadmap work. As action officers, Joseph Eckersley, Betty Spofford, Julie Altham, and Sheryl Davis (565 CBSS/GLBD) facilitated our contacts with all the ALCs and helped us find and acquire data. Florencio Garza and CMSgt Greg Laird (ACC/A4MA) also helped us with contacts and data at units and at Air Combat Command headquarters.

At Warner Robins ALC, we benefited from discussions with John Stabler (742 CBSG), Walter Blount and Dempsey Ventress (402 EMXSS/MXDEAC), Robert Pennington (581 SMXS), Rodney Selmen, and Geoffrey McGowan (402 EMXG/MXDEAD).

At Oklahoma City ALC, we met with Karen Hagar (533 ACSS/CL), Carlos Luna, Mark Quinlan, Chad Burks, Kathie Smith, and Lisa Jimenez. Janet Hulin, manager of the Automatic Depot Test Station program, provided us with information about the capabilities and transition issues of this new tester. Jerry Osborne (550 CMMXS/MXDPBA) provided key information on B-1 avionics repair.

We discussed B-1B avionics repair and ATS modernization issues in general with Ron Smith, Mike Clark, Al Lawrence, Bill Lee, and Jeff DeVries (309 EMXG) at Ogden ALC.

At Ellsworth Air Force Base, we acquired data and consistently good advice from MSgt Jerry Pierce and MSgt William Hanson (28 MXS/MXMV).

Special thanks to Bernard Smith and Woodrow Parrish (Logistics Management Institute, at the Air Force Logistics Management Agency) for help with the D200 data.

We also appreciated the comments and advice from Vernon Hilderbrand and Col Socrates Green at AF/A4LY, and from RAND colleagues Kristin Lynch, Raymond Pyles, Laura Baldwin, and Don Snyder.

Comments on an earlier draft of this report were provided by WR-ALC/GRN, the Air Force Logistics Management Agency, and ACC/A9M. We also appreciate the thoughtful comments and suggestions made by our RAND reviewers, Marc Robbins and Guy Weichenberg.

---

[1] The ranks and office symbols used here and in the text were current when the research was conducted. Since then, there have been a number of reorganizations and office symbol changes (e.g., the 742 CBSG is now WR-ALC/GRN).

Abbreviations

ACC	Air Combat Command
ACPINS	Automated Computer Program Identification Number System
ADTS	Automatic Depot Test Station
AFB	Air Force base
AIS D/I	avionics intermediate shop, display and indicator test station
ALC	air logistics center
ATE	automatic test equipment
ATLAS	Abbreviated Test Language for All Systems
ATS	automatic test system
CASS	Consolidated Automated Support System
CBSG	Combat Sustainment Group
DATSA	Depot Automated Test System for Avionics
DAV	digital analog video test station
DEM	repairs demanded per year
DIG	digital test station
DoD	Department of Defense
EASTE	enhanced automatic special test equipment
EPCAT	enhanced power control assembly tester
FAP	fleet application percentage
FEDLOG	federal logistics data
GAMS	General Algebraic Modeling System
GAO	Government Accountability Office (formerly General Accounting Office)
IFTE	Integrated Family of Test Equipment

ITA   interface test adapter

LRU   line replaceable unit

MDS   mission design series

MILP   mixed integer linear program

NATS   New ATS

NIIN   National Item Identification Number

NSN   national stock number

OIMDR   organizational intermediate maintenance demand rate

PAF   RAND Project AIR FORCE

PGM   product group manager

REW   Radar Electronic Warfare test station

SPO   system program office

SRU   shop replaceable unit

TPS   test program set

USTB   upgraded system test bench

UUT   unit under test

VDATS   Versatile Depot Automatic Test System

WATS   Web Automatic Test System

CHAPTER ONE

A Roadmap for Modernizing Air Force Automatic Test Systems

Nearly all the electronic and electrical components in Air Force aircraft and other equipment are tested and diagnosed using automatic test systems (ATSs), computer-controlled sets of instruments that generate inputs to the components, measure outputs, and identify problems. However, two related problems currently beset the Air Force's test systems: increasing equipment and software obsolescence and the difficulty of managing the many specialized types of equipment.

While piecemeal ATS modernization could help stave off obsolescence, it would be expensive and time-consuming because making new, modern instrumentation work with significantly older hardware is technically complex. On the other hand, buying completely new systems also has significant costs, such as for purchasing hardware and rewriting software. New purchases must also comply with current Department of Defense (DoD) and Air Force policies that direct replacement of mission design series (MDS)–specific equipment with standard testers that can test components from multiple weapon systems.

The modernization decision therefore requires balancing and trading off complex cost and capability information to decide which legacy systems to retire in favor of new ones and which to retain. The purpose of the work documented here was to develop a roadmap methodology to guide the transition of workloads from the Air Force's many legacy systems to modern DoD-approved common ATS families.

Terminology

An *ATS* consists of all the hardware and software required to test a set of components. It includes both automatic test equipment (ATE) and test program sets (TPSs). The *ATE* consists of the core tester hardware—a computer and its operating system and the set of digital and analog instruments the computer controls—that can generate test signals for the unit under test (UUT), record responses from it, and make diagnostic inferences. A TPS for a UUT has three parts:

1. **testing software**—The software, which is almost always specific to the UUT, contains instructions for a sequence of functional and/or diagnostic tests. These instructions control the test instruments, providing inputs to the UUT, then recording and analyzing the outputs from it to determine whether they are correct and, if not, the possible causes.

2. **an interface test adapter (ITA)**—Each UUT has its own set of connections and input and output ports. The ITA physically connects the ATE and the UUT, routing signals from the various instruments in the former to the appropriate input/output pins in the latter. Related UUTs may share ITAs when feasible.

3. **documentation**—These instructions tell the operator how to run the required tests and how to interpret the results. Documentation for legacy testers is primarily on paper, while that for more modern systems is electronic.

In general, a given ATS type can test tens to hundreds of UUTs by varying the TPS and ITAs.

Obsolescence

Under the Air Force's Integrated Weapon System Management concept, a system program office (SPO) acquires and supports not only the weapon system for which it is responsible but also the platform-specific test equipment for maintaining that system throughout its life cycle.[1] Every individual weapon system now in service was therefore procured along with test equipment designed and purchased specifically for that system and its components. However, as has been noted elsewhere (e.g., Keating and Dixon, 2003; Gebman, 2009; Ramey and Keating, 2009), many of the aircraft the Air Force flies today have been in service for decades, and much of their support equipment therefore contains older and obsolete technology that is reaching the end of its useful life. With age comes the most basic form of obsolescence: Components become increasingly unreliable and fail, decreasing ATS availability and therefore testing capability.

The obsolescence problem is more complex than just the increasing downtime involved in replacing parts that fail ever more often. Because much of the legacy test instrumentation technology is outdated, it is increasingly difficult or even impossible to get replacement parts to support the original testers. Original manufacturers may have stopped making the needed parts, and some may have left the business altogether. However, new components and instruments are technologically more advanced and can be incompatible with the technology of an existing ATS. Replacing the instruments on a legacy ATS can therefore require substantial engineering and software rework. Even without a technology gap, instrument replacement has problems due to the complexity of these systems. When any part of a tester changes, the interfaces between different parts of the system become vulnerable. Languages, hardware, and communication protocols must all be compatible, and the specifics of each test in the TPS must be calibrated to the test instrumentation. As a result, updating either the component being tested or any part of the test equipment may require changes throughout the system.

TPS software has its own set of obsolescence problems. Many legacy test programs were written in older test languages (e.g., the Abbreviated Test Language for All Systems, ATLAS) that are no longer widely supported. Much code was written before modern standards for software development; many of the original programs are not modular, not well documented, and difficult to understand in general. In addition, the TPS software, and sometimes the language, is proprietary in many cases. It has become expensive or impossible to continue to pay the original contractor for software updates, and it is extremely costly to reverse engineer the

---

[1] One ATS may support components from several platforms, with all the SPOs involved providing financing.

testing procedure without access to the original source code. Some contracts prohibit reverse engineering the code. Finally, as with hardware, some original legacy software contractors are no longer in business.

This procurement policy has also led to many different ATS types, particularly for avionics, to support a variety of Air Force systems even though the UUTs might have quite similar functions.[2] The large number of unique, incompatible ATS types across the Air Force complicates the problem of maintaining and updating aging systems. Each type has different parts, different software, and different instrumentation. Obsolescence fixes for one ATS type dedicated to one MDS provide little or no benefit to another type for similar UUTs for another MDS.

Further, because these testers are located at operational units and other repair facilities, such as Air Force Materiel Command's ALCs, and must be functional for the facilities to accomplish their repair missions, replacing legacy systems with new ones requires careful planning to avoid disrupting repair activities. Although the MDS SPOs are ultimately responsible for ATS support, Air Force units and the ALCs have sometimes had to take action themselves by using a variety of ad hoc methods to ensure the availability of legacy systems. The organizations cannibalize other testers for parts, make piecemeal and incremental updates, add translation code between old and new software, and add interface devices between old and new and hardware to carry signals. Despite these efforts, testers in some locations are down for substantial periods.

In summary, keeping an obsolescent ATS running requires continuous management to address emerging problems and devise ongoing strategies.

Common Automatic Test Systems in the Department of Defense

The problem of aging test systems has not been limited to the Air Force. Up to the mid-1990s, different, incompatible ATS types were proliferating throughout DoD. However, software engineering and computer-controlled hardware have matured considerably since the first such systems were developed over 40 years ago, and it is now standard practice to modularize both hardware and software using internationally accepted standards for communication protocols within and between hardware and software modules. The new open-architecture approach for interconnecting electronic systems has made it possible to develop ATS families with standard architectures so that replacing one piece of a system does not necessarily also entail significant redesign of the whole. This facilitates a move away from a collection of proprietary and legacy testers with a wide variety of incompatible hardware and software to a smaller set of ATS families that can test many components from different systems. In the early 1990s, DoD made it policy to acquire all ATE hardware and software from designated families and to design new testers as open systems. The intention was for each ATS to be able to test as many different components from as many different weapon systems as possible and thus to end up with the smallest feasible set of interoperable ATS families (Under Secretary of Defense for Acquisition

---

2 Personnel from the 742 Combat Sustainment Group (CBSG) at Warner Robins Air Logistics Center (ALC) quoted numbers in the range of 300 different types in the Air Force. They noted that Warner Robins ALC alone had 260 different testers. Meeting with 742 CBSG on December 16, 2008.

and Technology, 1994).[3] The new policy was included in the 1995 version of DoD Instruction 5000.2.

The ATS policy was removed from DoD Instruction 5000.2-R in 2001; however, in response to a 2003 U.S. General Accounting Office (GAO; now Government Accountability Office) report that emphasized the need to better manage ATS modernization (GAO, 2003), the original policy was reestablished in a memorandum from the Office of the Under Secretary of Defense for Acquisition, Technology, and Logistics (Wynne, 2004). The memorandum provided specific guidance to program managers concerning ATS modernization. The goals of the policy are to minimize life-cycle cost in manufacturing operations at all maintenance levels, promote joint ATS interoperability, and minimize unique ATS types in DoD.

Current DoD ATS policy directs the services to develop a small number of ATS families. Each family is to be interoperable with commercial instruments and communications buses, support a variety of weapon system requirements through flexible hardware and software architectures, and be expandable and customizable without requiring basic architectural changes. The ATS Framework Working Group develops the elements each family architecture must include, and as each element is tested and approved, it becomes a mandate. The basic architecture requirement is that the ATE's internal hardware, instrumentation, drivers, and associated software be independent of the individual equipment being tested and use commercial-off-the-shelf components wherever practical. By 2004, DoD had approved four designated ATS families. The Army, Navy, and Marine Corps each had its own versions, along with the Joint Service Electronic Combat Systems Tester.

Motivated by the space constraints of shipboard repair facilities, the Navy developed the Consolidated Automated Support System (CASS) as its first single family of testers capable of testing across multiple platforms.[4] CASS was first designed in 1986, ordered in 1990, and entered service in 1994. Its basic core test station provides analog and digital capabilities, and other versions add capabilities for specific testing requirements (radio-frequency components; high-power radar systems; electro-optics; and communications, navigation, and identification friend-or-foe systems). The introduction of CASS reduced the kinds of training and specialties required because all forms of CASS share a basic core system, and a reduction in manning requirements followed.

Even as a common tester family, CASS has already undergone updating and modernization. The original CASS was encountering obsolescence issues by 2006, when many of the test stations had seen over 100,000 hours of use. Frequent replacement of worn-out physical components increased station downtime. Original CASS software had a closed architecture, which was inflexible and hard to update. The newer Electronic CASS is designed to take advantage of advances in state-of-the-art system software, new TPS and other programming languages, operating systems, and bus architectures. The Navy's ATS acquisition strategy is to build around this system as the standard family. Naval Air Systems Command PMA260 centrally manages these acquisitions and is also the ATS executive directorate for DoD.

The Army faced analogous constraints, in that its testers have to be easy to move and ruggedized for field use. The Army's standard ATS family is the Integrated Family of Test

---

[3] *Open systems* refers to the concept of highly interoperable computer infrastructure based on nonproprietary standards for software and hardware interfaces (Open Group, 2011). For history and current details of DoD policy on common ATS, see the ATS Executive Agent's home page, 2011.

[4] For a description of CASS, see Naval Air Systems Command, undated.

Equipment (IFTE), which includes both at-platform and off-platform diagnostic test systems. At-platform testers allow soldiers to diagnose and fix equipment as it breaks in the field. There are several versions of at-platform testers, including the Soldier Portable on-System Repair Tool. Off-platform IFTE likewise comes in several versions designed to support specific testing needs, such as electro-optics. Although the Army currently maintains several system-specific testing families, the Next-Generation ATS (Burden et al., 2005), the latest test station in the IFTE family, is intended to become the Army's one common core tester. It will be able to test components from all weapon systems and will be backward-compatible with current systems. U.S. Army Materiel Command's Test, Measurement, and Diagnostics Equipment program centrally manages the service's ATS acquisition.

Managing ATS Obsolescence and Commonality in the Air Force

Air Force constraints differ from those of the Army and Navy. The introduction of the aerospace expeditionary force concept and centralized repair facilities in 1998 lessened the need for common testers small enough to be deployed with a unit to its operating location, ameliorating one key driver that pushed the Navy and the Army to move to CASS and IFTE (see, e.g., Geller et al., 2004). This lack of the binding constraints the other services faced, combined with the Air Force tradition of procuring system-specific testers, continued ATS proliferation into the early 2000s.

The Air Force too, however, has multiple testers heading for obsolescence and faces the new DoD policies. It has responded by taking steps to centralize ATS management and to establish Air Force families of common core testers. The first such DoD-approved Air Force family is the Versatile Depot Automatic Test System (VDATS), which contains three interoperable testers that are intended to be capable of testing virtually all avionics line replaceable units (LRUs) and shop replaceable units (SRUs) for all MDSs. In addition, individual weapon system SPOs have developed and fielded other modernized testers, such as the Automatic Depot Test Station (ADTS), which replaced two aging B-1B avionics testers in the Intermediate ATE suite, the digital test station (DIG) and the digital analog video test station (DAV).

Like the other services, the Air Force has now centralized its management by assigning the ATS Product Group Manager (PGM) at 742 CBSG at Robins Air Force Base (AFB) as the single manager and leadership office for ATS. The ATS PGM manages VDATS and other testers, establishes ATS solutions for acquisitions and sustainment support to SPOs, and manages the movement of all Air Force testing to common core testers.

Initial ATS PGM efforts to define the scope of the problem have been complicated by the fact that the ALCs and SPOs have pursued their own solutions independently, designing and contracting for ATS upgrades and replacements. In addition, a comprehensive picture of the status and obsolescence problems of the Air Force's testers has been difficult to assemble because of fragmented data systems, the sheer numbers involved, and the lack of previous centralized management.

A Roadmap for Air Force ATS Modernization

As noted previously, all the services must replace their legacy systems with the DoD-approved ATS families, as feasible. The five DoD-designated families may evolve or change, but the number of families is expected to remain small. As the Air Force plans the transition to new ATS families, it faces the problem of deciding which existing legacy testers to replace before their life cycles end and in what order. It is infeasible, both economically and operationally, to migrate all the workloads for all the testers simultaneously. Air Force weapon systems contain hundreds of electronics LRUs and thousands of electronics SRUs.[5] Estimates for rehosting TPSs on common core testers vary, but even an optimistic estimate of $500,000 per LRU and $150,000 per SRU would put the total cost of rehosting in the billions of dollars. The time involved is also daunting: It can take as long as one and a half years to rehost a single TPS for a complex LRU. Finally, not all components need to be rehosted. There are plans to retire certain platforms in the near future, which would eliminate testing for many of their components.

A migration schedule has to take a number of factors into account, including the urgency of operational requirements, the technical feasibility of transition, and the cost-effectiveness of the transition given the projected life cycle of the supported MDS and the ATS itself. Other factors may come into play in evaluating trade-offs: planned upgrades to the supported systems, the reliability of the existing testers, planned expenditures to purchase new testers or replace existing testers, differences in tester performance, the effects on fleet availability, and available space for new equipment.

Research Scope and Report Organization

The Air Force needs to plan for the eventual migration or retirement of all test workloads for all weapon systems and their associated support equipment. The purpose of our work was to formulate a general methodology for the economic analysis, by UUT workload, of migration from legacy to new testers and to apply that methodology to a test case workload. It did not address the engineering or design issues involved in producing testers in the new family and does not seek to validate or endorse a particular family of testers. Currently, VDATS is the only Air Force–specific designated ATS family, but VDATS may evolve or be joined or replaced by another family of universal testers, leading to more than one DoD-approved ATS family in the Air Force.

Chapter Two describes the decision space for constructing a roadmap. In Chapter Three, we offer a mixed integer linear program (MILP) for solving the UUT rehost problem and define the data we need for that program. We also discuss some of the data problems we encountered, which will affect future Air Force ATS management efforts. In Chapter Four, we apply the methodology to a subset of B-1B avionics UUTs and testers and examine the recommendations the roadmap methodology makes under a variety of assumptions and parameter values. Chapter Five draws general conclusions about the UUT rehost decision. The appendixes contain supporting material.

---

[5]   SRUs are designed to be tested and replaced in avionics shops or at an ALC. They are components of the more complex LRUs, which are designed to be replaced on the flightline.

The Rehosting Roadmap Decision Problem

Components of the Decision Problem

The purpose of the rehosting decision is to identify which maintenance workloads to shift from legacy testers to new, modern testers and to lay out the rehosting sequence (the *roadmap*). We focused on the economic aspect of the rehosting decision, i.e., what should be rehosted from legacy testers to new testers to most efficiently use resources while maintaining repair capabilities. In this chapter, we describe the individual components of the decision problem, which will inform our choice of solution methodology and determine which data must be collected to construct the roadmap.

Costs of Legacy Automatic Test Systems

The cost of continuing to operate legacy testers is the amount needed to keep them sufficiently available to handle the required workload. This cost has three components:[1]

1. *per-tester operating costs* for running the tester, such as electricity and manpower
2. *per-tester maintenance costs* to keep individual testers running
3. *per-type sustainment costs* to keep the entire ATS set of a given type operational.

The second component covers such activities as repair and calibration. The third component, besides maintaining spares stock, technical orders, and requiring management resources, may include finding new sources of components because legacy components and instrumentation may no longer be manufactured. If replacement parts or instruments are not available, support personnel must reengineer and/or modify the legacy ATS to accept new test instruments. This modification can require both electronic work and rewriting the TPS.[2]

Costs of New Automatic Test Systems

Acquiring a new tester also has substantial costs:

1. acquisition of new tester hardware
2. rehosting existing TPS on the new hardware
3. building new hardware interfaces to connect the new tester to the UUTs.

---

[1] There are additional costs for program support, but we do not include those because they are hard to determine and would also apply to some extent to any new tester.

[2] Discussion with personnel at Warner Robins ALC, May 18, 2009.

The new equipment will also have its own operating, maintenance, and sustainment costs (although these are usually assumed to be lower than those for legacy systems because the technology is new and because common testers offer economies of scale).

Time Horizon

The roadmap has an inherent temporal component. For any particular MDS, the repair workload has (in principle) a finite time horizon. This stems both from the potential aircraft retirement and from component and system upgrades, resulting in the retirement of older LRUs and SRUs.

For this reason, the length of the roadmap will influence rehosting decisions. TPS translation costs should be expended on UUTs that will be in the inventory long enough to get some return on the rehosting expenditures. This requires comparing the costs of supporting a legacy ATS with the cost of rehosting its UUTs to a new tester by discounting both costs over the time horizon and comparing the net present value using an appropriate discount rate.[3]

Decision Space

Our main focus will be on deciding which UUTs to rehost on a single, new, common ATS. Much of the discussion implicitly assumes that the decision would be made at the level of the legacy ATS type, i.e., whether a tester of a given type should be retained at full capacity or whether all the UUTs repaired by that type should be rehosted. Deciding between these two alternatives makes sense for the Navy and to some extent the Army, which have important space and bulk constraints on much of their testers. However, as noted previously, the Air Force does not, in general, have these physical constraints.

We will therefore use a finer, more granular decision space, deciding UUT by UUT which ones to rehost. This level of granularity includes both the other alternatives as special cases (no UUTs move or all move) and also makes our decision more flexible. For example, if the workload for a particular legacy ATS type is dominated by a single UUT, but the ATS also tests many others with low failure rates, it may be more efficient to translate only the TPS for the high-demand item and use the now substantially idled remainder of the ATS type for cannibalization because fewer of them would be needed for the remaining repairs. This would be especially attractive if the low-demand UUTs will be phasing out within a few years.

Issues Not Considered

Switching to a new, common tester has several advantages that could offset the costs of rehosting. However, quantifying these benefits is much more difficult than assigning costs for legacy ATS support or UUT rehosting. Furthermore, any savings from these sources would likely be spread out over future years near the end or beyond the ten-year horizon we used for the case study in Chapter Four. Our analysis has therefore *not* considered the following issues:

---

[3]   Net present value is the value today of a set of future expenditures, discounted by their distance in the future. This is a standard method of comparing future income or expenses. See, e.g., Higgins, 2007.

- A new, common tester should be easier to maintain than multiple testers. Spare parts would be common, and maintainers would all be working on a much more limited ATS set.
- Repair technicians should be more flexible. Working on a common ATS should make it easier for technicians to cross-train to work on different workloads, as needed.[4]
- Repair facilities can become more efficient. Because it takes fewer common testers than specialized equipment to handle a diverse workload, the demands for space and power may decrease. This also means that workloads can be cross-leveled much more easily across different MDSs and ALCs.
- Repairs may be more accurate and comprehensive because the more-capable instrumentation has better diagnostic capabilities.[5]
- A more homogeneous ATS set should require fewer logistics and management resources (e.g., fewer spare parts, less contracting).

---

[4] When the Navy replaced a set of diverse legacy ATSs with CASS, it was able to reduce the number of occupational specialties in the avionics shop from 32 to 4 and the number of ATS operators from 105 to 54 (GAO, 2003, p. 9).

[5] In our analysis, we made the assumption that repair times would not change with a new tester, because the UUTs have older technology and may not be able to be tested more quickly. This may not be uniformly true, however. Also, new testers can have better and more-integrated maintenance documentation, which could also help cut repair times.

CHAPTER THREE

Methodology and Data for the Roadmap

Overview of the Methodology

The previous chapter laid out the components of the roadmap decisions and potential options. For each UUT, the roadmap should provide a decision about whether to rehost that workload on a new tester or continue to use the legacy tester for some time. If that decision is to rehost, the roadmap must specify how many new testers will be needed to handle the rehosted workload and must include support cost adjustments for the legacy systems that will still be in use. To select an optimal strategy of discrete actions based on the costs and constraints, we formulated an MILP.[1] The next section describes this program in more detail, while the subsequent sections describe how we assembled the data required to implement the MILP.

The Roadmap as a Mixed Integer Linear Program

The salient issue in making the economic decision to replace an old ATS with a new one is whether the projected savings in operating the new tester rather than the legacy ATS over the remaining life of the supported weapon system will outweigh or "pay back" the initial investment in hardware and software that must be made to field the new testers. However, as we noted above, it is neither necessary nor optimal to take an all-or-nothing approach by MDS or by ATS type. The workload on a given legacy tester can be partitioned to look at each unique UUT as an independent rehost–no rehost opportunity. That is, some of a given legacy tester's UUTs might go to a new system, while the rest might remain on the old tester. And to be more precise, the decision to rehost a UUT is not binary, either. That is, it is possible to rehost part of a given UUT workload and leave the remainder on a legacy tester. Thus, the number of possible or feasible solutions to this rehosting decision is very large, and we need an optimization algorithm to identify the best possible solutions.

We therefore developed an MILP using the General Algebraic Modeling System (GAMS) to develop cost-optimal solutions to the ATS replacement problem.[2] This model searches all possible replacement decisions to find the cost-optimal decision. This model formulation also facilitates sensitivity analyses that point out which parameters (which costs, for example) may drive the solution results and which are relatively unimportant.

---

[1]   For an introduction to MILPs, see any basic operational research book, e.g., Hillier and Lieberman, 1995.

[2]   GAMS is a high-level modeling system for mathematical programming and optimization. It is available from the GAMS Development Corporation website.

The MILP model also includes the ability to distinguish between three different kinds of costs associated with operating and maintaining both legacy and replacement testers. These are the costs associated with

1. **tester use**—For example, while a given UUT is in work, labor hours, electrical power, etc., are consumed. Some maintenance costs are also a result of usage. In the model, this *usage cost* is based on the total tester time planned for a given tester type during a given year.
2. **number of testers of a given type**—Regardless of use or utilization, this equipment needs to be maintained. For example, a tester may need to be calibrated and serviced every six months regardless of throughput or usage. In the model, this *per-tester* cost is based on the total testers of a given type planned to be in the inventory in a given year.
3. **existence of a given type of tester**—Regardless of the quantity in inventory or frequency of use, a given type will need some amount of annual sustainment engineering as long as any of its testers remain in inventory. In the model, this *per-type* cost is charged for any tester type for any year in which the number of testers in the inventory is greater than zero.

The other costs in the model represent the rehosting costs and have two components:

1. the acquisition cost of a new tester—the cost to acquire one new tester of a given type in a given year, which is charged whenever a decision plan adds a new tester to the inventory
2. a specific rehosting cost for each UUT—the cost to translate or reprogram the test program software from a legacy tester to a new one, a one-time cost charged in the first year in which a UUT is rehosted.

Our model also incorporates other important aspects of the problem. Since the projected life cycle of a weapon system extends decades into the future, we used a multiperiod model to include the time dimension in the decision. This also allowed us to correctly calculate the net present value of expenses that occur across the planning horizon. For each of the three basic types of operating and maintenance costs, cost estimates could be provided for each tester type for each year in the life cycle or planning horizon so that we could model the expected growth in maintenance costs for legacy testers due to obsolescence and diminished manufacturing sources. In the model, rehosting costs can also have different values in different model years.

Our MILP has additional flexibility that we did not use for our example analysis. For example, the model can allow a decision to rehost any portion of a UUT workload at any point (year) in the remaining lifetime of the weapon system. This would allow constraining the maximum budget amount that could be spent on rehosting investments in a given year. It can also impose area constraints if repair space is limited.

In summary, then, we visualized the rehosting problem as an assignment problem. This assignment of workload is modeled at a high level of detail. The model directs the assignment of (potentially fractional) UUT workloads, year by year, between legacy and replacement testers to minimize the total operating and maintenance costs (of three types) and the total rehosting costs (hardware and software).[3]

---

[3] See Appendix A for details on the model formulation.

Data for the MILP

As noted previously, three basic sets of data provide the parameters for the MILP formulated above: the projected UUT workload, the cost to use and maintain both the legacy testers and the new ones, and the cost of rehosting UUTs on the new tester. We go into some detail about our data sources for the workload projection in this section for two reasons:

1. Assembling the data for the MILP was the most time-consuming part of the project, consuming more than 50 percent of our total effort. As we will relate, this was due to fragmentary and decentralized databases that use multiple ATS and UUT identifiers, which, in some cases, are not standardized.
2. Future ATS management will require repeated, continual access to reliable data of the type we assembled for the case study analysis detailed in Chapter Four. It was important to lay out our sources and difficulties as guidelines for making the data consistent and easily available in the future.

We will defer discussion of the cost data to the following chapter because it is more specialized to the legacy testers and UUTs we chose to illustrate our methodology.

Projecting the UUT Workload

Projecting the UUT workload for a legacy and a new tester requires two separate sets of data. The first is on the repair demand, i.e., the expected number of UUTs to be sent for testing and repair on the legacy ATS each year. The second is the match of UUTs to that system and the time it takes to test and repair each UUT.

UUT Demands

Computing UUT demands for the legacy ATS, we need a list of UUTs tested on them (both LRUs and SRUs), the UUT failure rates (usually expressed in removals per flying hour), the number of UUTs on an aircraft, the flying-hour program, and the portion of the aircraft fleet that carries the UUT.

An initial source of LRU and SRU data is Air Force Materiel Command's Requirements Management System database, D200. A list of LRUs and SRUs for a given MDS is relatively easy to obtain; however, identifying avionics components is more difficult. Work unit codes provide some indication of avionics function but are not always complete or accurate. Initially, we used an extract for the 2006 D200 but supplemented that with the latest 2009 data.[4]

D200 is the primary source of failure rate data for UUTs, providing a removal rate for each UUT scaled by flying hours for the aircraft fleet. This is the organizational intermediate maintenance demand rate (OIMDR), and most avionics equipment repair estimations are done using the OIMDR and projected flying-hour programs.[5] The other parameters are the quantity per application (QPA, the number of each type of UUT on the aircraft) and fleet

---

[4] For a mature MDS, such as the B-1B, parts do not change that quickly. We initially used the 2006 data while acquiring the 2009 update. We used the 2009 data to check and change data as needed.

[5] A number of UUTs that had repair records at the base or depot had 0 OIMDR. Some of these may be specially managed by item managers who calculate estimated requirements using other methods. They zero out the OIMDR to prevent conflicting demand estimates.

application percentage (the portion of the MDS fleet that has a particular type of UUT[6]). Since the OIMDR equates to the removals per flying hour for *each* UUT on an aircraft, the expected number of removals for UUT_i in a year, R_i, is estimated by

$$R_i = AFH \times OIMDR \times QPA \times FAP,$$

where *AFH* is the annual flying hours for the fleet and *FAP* is the fleet application percentage. The expected workload, *W*, for a tester type per year is

$$W = \sum_i R_i T_i,$$

where T_i is the repair time for UUT_i on the tester.

With the specification of a flying-hour program, an expected projected workload can be computed for each UUT.

UUT/ATS Match and UUT Repair Times

Unfortunately, the data to match UUTs to an ATS are not centrally located. Technical orders could, in principle, be primary data sources for matching, since these describe how to test a UUT on the applicable ATS. However, extracting the data from the technical orders for *all* ATS-related UUTs is daunting, since many of the orders are still available only on paper. Local repair shops obviously know which ATSs repair which UUTs, but this information is kept in formats that vary from place to place.

The Automated Computer Program Identification Number System (ACPINS) database (ACPINS, undated; U.S. Air Force, 2003a; U.S. Air Force, 2003b), which tracks computer programs by their identification numbers, came closest to providing UUT-to-tester information for all major Air Force aircraft. ACPINS is supposed to catalog every TPS that Air Force maintenance uses and link it to the UUT, the ATS, and the hardware interface. Unfortunately, many of the last group are referenced by part number, not National Item Identification Number (NIIN).[7] The crosswalks between part number and NIIN have to be resolved from some other data source, such as federal logistics data (FEDLOG),[8] as do the variations in part numbers (e.g., punctuation may vary, AN-35/G is the same unit as AN35G). Therefore, ACPINS data had to be substantially refined and complemented with data from several other data systems to make them usable for our ATS analysis.[9]

There was a further complication. ATS equipment is typically identified in one of a variety of ways, such as by an acronym, an Army-Navy equipment designator (e.g., AN/GSM-305(V)5), a part number, a national stock number (NSN), a standard reporting designator, or even a manufacturer's model number. However, some equipment does not have an acronym,

---

[6] This percentage is important because some MDS fleets have different variants that may not have a particular system with its associated UUTs.

[7] In many cases, the part numbers were current when the TPS was introduced. However, the number itself may include the AN-designator of the UUT or the ATS tester.

[8] For information on FEDLOG and the Federal Logistics Information System, see Defense Logistics Agency, 2011.

[9] Some B-1B workload at Oklahoma City ALC is being moved to new testers, ADTS and the enhanced power control assembly tester (EPCAT), so tester assignment may be different in different data sets.

AN-designator, and NSN (see Table 3.1). Furthermore, a given acronym may refer to different testers (e.g., there are various power-supply test sets). Worse yet, cross-reference tables relating one identifier to the others are not centrally available, and the mappings may not be one-to-one (each ATS may serve multiple NSNs and part numbers).

UUT repair times are also kept in various forms and in various ways at different repair facilities. For example, each ALC uses a different data system for maintenance data collection.

Development of a Data Model

Using these data sources in parallel and resolving ambiguities by cross-checking and hand editing, we built a preliminary integrated relational database of UUTs and their testers for all major Air Force aircraft. Table 3.2 shows our data sources. All the data were current at the time of our research, except where noted. Data limitations were mostly inherited from the data sources.

To store this information, we created a dedicated database (Web Automatic Test Systems, WATS), which combines a web-based interface for queries with static database snapshots from our data sources. We used these snapshots for the case study described in the next chapter, but WATS actually contains data from a wide variety of MDSs. If the data for WATS were continually updated from the data sources we identified, it could be a valuable tool for ATS management.[10]

Table 3.1
Sample ATS Identifiers

ATS Acronym	Aircraft	AN Designator	Sample NIIN	Sample Part Number	Standard Reporting Designator
VDATS-DA1	Various		015530693	200625630-10	
DATSA	B-1, etc.	GSM-305V5	012031835	865300-305	GLH
ATS	F-15	GSM-228	013870979	13A6520-5	
REW	B-1		011783451	3200030-118	
AIS D/I	F-16		013205388	2212350-002	GSY
EASTE	B-1, etc.	ALM-280	013452122	SK1200-3	G5V

---

[10] See Appendix C for more detail on WATS.

Table 3.2
ATS Data Sources

Data Source	Scope	Coverage UUTs	Coverage UUT to ATS	Coverage ATS Tester
Air Force ACPIN	Air Force–wide	Part numbers	Yes	Many part numbers
Depot B-1 avionics shop	Oklahoma City ALC	Part numbers, most NSNs	Yes	Acronym[a]
Base B-1 avionics shop	Ellsworth AFB	Part numbers, most NSNs	Yes	Acronym
Depot electronic warfare shop	Warner Robins ALC	Part numbers, most NSNs	Yes	Acronym
Lean Depot Management System	Warner Robins ALC	Part numbers, many NSNs	Yes	Part numbers, many NSNs
Defense Repair Information Logistics System	Ogden ALC's F-16	NSNs, part numbers	Yes	Acronym
Scheduling and Kitting Inventory Listing[b]	Oklahoma City ALC	?	?	?
Discoverer	Air Force–wide	NSNs, part numbers		NSNs, part numbers
FEDLOG data service	DoD	NSNs, part numbers		NSNs, part numbers
Federal Logistics Information System	DoD	NSNs, part numbers		NSNs, part numbers
Logistics On-Line Classic	DoD	NSNs, part numbers		NSNs, part numbers
D200	Air Force–wide	NSNs		NSNs
Air Force Calibration Authority Viewer	Air Force–wide			Part numbers
Precision Measurement Equipment Laboratory Automated Management System	Not depots			Part numbers
Facilities Equipment Management System[b]	One per depot			Part numbers
Air Force Equipment Management System (2006–2007)	Air Force–wide			NSNs
Test Equipment Database	Warner Robins ALC			Many part numbers

[a] *Acronym* indicates that little corresponding information was available, e.g., mainly just the tester acronym. Defense Repair Information Logistics System, undated.

[b] We did not get access to these data systems.

CHAPTER FOUR

Case Study: Selected B-1B Avionics

Rationale for B-1B Avionics Case Study

To create an example for the roadmap methodology, we selected an ATS set used to repair avionics on the B-1B. This set has a number of attractive attributes for applying our methodology (Air Combat Command [ACC], 2001):

- The ATS set for B-1B avionics is of current concern, with some testers having substantial downtime and requiring significant efforts to find or replace obsolescent parts.
- Avionics components that are not repaired on base are repaired primarily at Oklahoma City ALC. Some electronic warfare components are repaired at Warner Robins ALC, but only a handful at Ogden ALC. The concentration of repair activity at Oklahoma City ALC eased data collection on characteristics of the repair process.
- The B-1B's fairly small fleet (93 aircraft) is concentrated at two bases: Dyess AFB and Ellsworth AFB. As with the previous attribute, this simplified data collection.[1]
- The fleet is homogeneous, with virtually all avionics components used on all aircraft. This means that we did not have to collect detailed fleet application percentages on subsets of the fleet.

These characteristics suggested that data acquisition and other information gathering could focus on a small number of organizations. However, this was not a comprehensive analysis of the entire B-1B avionics suite. We focused our attention on a well-defined set of testers and UUTs, as described below, and executed our methodology on these data. As will be obvious, the same methodology could be applied to a single ATS type, to the ATSs for several different MDSs simultaneously, or potentially to all ATSs the Air Force uses for aircraft and other equipment (although the data collection effort would likely be very challenging). Although the data can be improved (and we will make some suggestions to address this issue), we believe that the findings delineate the broad structure of an ATS roadmap for the Air Force.

We will compare the legacy testers with a notional new tester, which we will designate as *NATS* for "New ATS." It has the ability to test and repair all the UUTs from the legacy ATS and has cost and availability similar to VDATS.

---

[1] Federation of American Scientists, undated. This number may include more than just the aircraft in actual use, but this is not relevant for the case study.

Data and Assumptions

Figure 4.1 illustrates the data requirements and outputs for our MILP methodology. We will discuss the details of each of the inputs for the B-1B example in the following sections.

B-1B Avionics: ATS Included

Our analysis included the following current ATS equipment:

- The Intermediate Automatic Test Equipment suite, which was based on F-16 test stations from the 1970s and entered service in 1984. The suite comprised three test sets:
  - the Radar Electronic Warfare (REW) test station
  - DIG
  - DAV.
- The AN/GSM-305 Depot Automated Test System for Avionics (DATSA). This ATS has supported both the B-1B and the F-15 and dates from the late 1970s. Its primary function is to test SRUs, usually circuit cards.
- EPCAT is used primarily on the B-1B and was introduced in the early 2000s.
- The upgraded system test bench (USTB) has taken over some of the workload from the older REW from the Intermediate Automatic Test Equipment suite.[2]

Although we refer to these six as "legacy" testers, the EPCAT and USTB are of much more recent vintage than the other four and are substantially more reliable and easier to main-

Figure 4.1
MILP Input Data and Output Measures

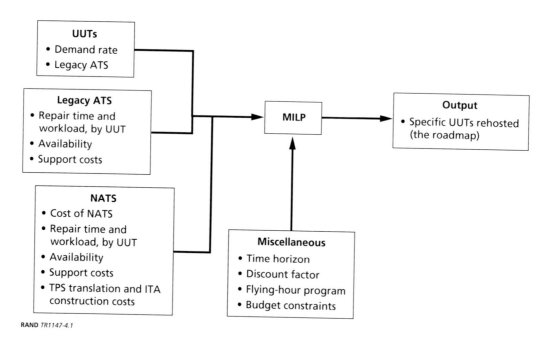

RAND TR1147-4.1

---

[2]   Technically, the USTB is not considered an ATS by the ATS PGM but is a "hot mockup." But since it could, in principle, be replaced by a common ATS we included it in the case study.

tain. We included them here to add variation in ATS availability and support costs that would help us evaluate how our methodology works with an ATS mix.

The DIG and the DAV are in the process of being replaced by ADTS, which is under development for B-1B repairs at Oklahoma City ALC. ADTS is already in operation at Oklahoma City ALC, Dyess AFB, and Ellsworth AFB, and the workload is being steadily rehosted to the new tester. As with the EPCAT and USTB, we included the older ATS combination to illustrate our methodology.

B-1B Avionics: UUTs

Using the sources and methods just discussed, we assembled a list of UUTs tested on the six legacy systems. We relied primarily on UUTs identified as B-1B avionics in the Air Force's D200 data system, supplemented by UUT lists from the B-1B avionics shops at Ellsworth AFB, Oklahoma City ALC, and Warner Robins ALC and the Lean Depot Management System for tracking repair data at Warner Robins ALC. We excluded parts repaired at Ogden ALC because of their low number.

This combination of sources resulted in approximately 1,100 unique NIINs. We then selected components tested and repaired on our legacy ATS, again using information from our three sources of data.[3] Finally, we restricted the UUTs to components that had a nonzero removal rate in D200 and that had nonzero repair times in at least one of the three sets of repair data that we acquired from the three sources of repair (Oklahoma City ALC, Warner Robins ALC, and Ellsworth AFB). If there were multiple nonzero repair times, we used the longest one. This process left us with 470 UUTs. The SRU/LRU classification from D200 was used, except that all parts that were tested on the DATSA were classified as SRUs.[4]

Workload Projections

Chapter Three described our method for computing the annual workload for each ATS type. To illustrate our methodology, we assumed a level flying-hour program for the B-1B fleet of 2,000 hours per month (Office of the Secretary of Defense, 2008).[5] After computing the expected annual demands for each UUT, we computed the annual workload for each ATS by summing the annual workload across all the UUTs tested on that ATS. Figure 4.2 shows the distribution of UUTs and workload for each legacy ATS.

Each bar in Figure 4.2 represents the total annual workload in hours computed for each legacy ATS type. The number at the top of each bar gives the total number of UUTs for that ATS according to our data. Each bar is subdivided into the aggregate repair times for individual UUT types; the UUT with the largest work share for each ATS is at the top of the bar, followed by the second largest, third, etc. While the total DATSA workload is moderate, this

---

[3] Some components are actually variants of a single type and can be used in place of each other. For our purposes, we identified these sets of components by noting which had a common TPS (based on the identification number of the TPS). We then used a common NIIN for these components (usually the "subgroup master NIIN"). Rolling up NIINs with a common test program is necessary because rehosting costs are based on translating TPS and building interfaces; for items with a common TPS, the translation has to be done only once. Treating these as separate UUTs would overstate the rehosting costs.

[4] Appendix B provides a complete list of the UUTs we used.

[5] Analysis of Reliability and Maintainability Information System and Comprehensive Engine Management System data from B-1B operations.

ATS repairs 350 of the 470 UUTs in our data, and most of these have small individual work-loads. In contrast, USTB handles only four out of 470 UUTs, but it has a larger workload than DATSA, and the workloads of two of its four UUTs are themselves quite large. There is great diversity among the six testers in terms of number of UUTs, UUT workloads, and total work-load, which had a strong effect on the roadmap we computed.[6]

Legacy and NATS Availability

Table 4.1 lists our data on availability. Legacy ATS availability was derived from two sources. First, we gathered information from Ellsworth AFB about the availability of its test equip-ment (DIG, REW, DAV, EPCAT, and USTB). For these, the available time excluded sched-

Figure 4.2
Total and by-Unit-Under-Test Workloads for B-1B Legacy Automatic Test Systems

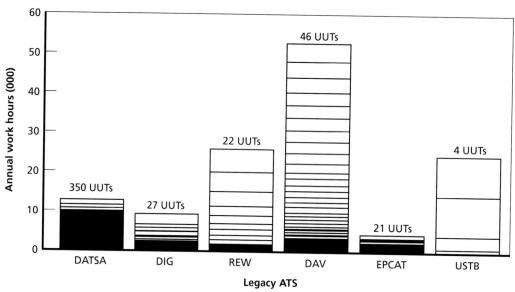

RAND *TR1147-4.2*

Table 4.1
ATS Availability

ATS	Availability (%)	Operating Hours per Year
DATSA	70	2,934
DIG	85	3,542
REW	48	1,993
DAV	84	3,502
EPCAT	95	3,952
USTB	93	3,866
NATS	95	3,952

---

[6] We emphasize that the 470 UUTs making up the workload we analyzed with our methodology are only a selected subset of the LRUs and SRUs in the B-1B avionics suite and that some of the ATSs, notably the DATSA, repair UUTs other than those in our study, some from other platforms.

uled maintenance and calibration and average unscheduled maintenance. Since the Ellsworth avionics shop does not have a DATSA, we computed its availability using data from Warner Robins' test equipment database. For the availability of the NATS test station, we assumed 95 percent, based on both experience with VDATS and the newer USTB and EPCAT. We used these availabilities and assumed a nominal two-shift, five-day workweek (no holidays) to compute available hours per year per unit per ATS.[7]

Operating and Sustainment Costs for ATSs

Making the case to replace a legacy ATS requires current data on the time and money spent in operating, maintaining, and sustaining the equipment.

The hourly operating costs are usually specified in ATS cost analyses but are very small. For hourly operating costs, we used the hourly pay for two operators at $100,000 per year for each individual ATS in operation.[8] We also added an electricity cost of $2,600 per year for the legacy systems and $1,300 per year for NATS (Eckersley, 2008). As we will show next in our discussion of sustainment costs, even significant changes in electricity costs would not increase operating costs enough to play a significant role in our analysis.

Unlike the operating cost data, which are fairly straightforward to estimate, the per-tester maintenance and per-type sustainment costs for legacy ATSs are uncertain. This is a bit surprising, given that the perceived continual increase in these costs is one of the central justifications for replacing these testers. However, much of this work is organic to operational units and the ALC, and so separating manpower and other costs for specifically ATS maintenance and support is difficult. We used a combination of interviews and data requests at Ellsworth AFB and Warner Robins ALC to assemble a baseline set of costs. Because these data are uncertain, our analysis focused on varying baseline costs over a range spanning the most likely costs.

For per-tester maintenance costs, we had information based on our analysis of Ellsworth data for all but the DATSA (Table 4.2). The base costs are the original per-tester estimates. For

Table 4.2
Per-Tester Support Costs for Legacy ATS ($000)

Tester	Baseline	3	5	10
DIG	7.5	22.5	37.5	75.0
REW	37.0	111.0	185.0	370.0
DAV	7.5	22.5	37.5	75.0
EPCAT	7.5	22.5	37.5	75.0
USTB	7.5	22.5	37.5	75.0
DATSA	25.0	75.0	125.0	250.0
Range	7.5–37.0	22.5–111.0	37.5–185.0	75.0–370.0

[7] Ellsworth AFB has a higher repair tempo than this.

[8] Our interviews at the repair centers indicated that two operators are required for safety reasons with electrical equipment. Some ATSs are isolated, but others are in bays with multiple repair stations in operation. In the latter case, two operators may not be required. This is a conservative estimate. See Dahlman, 2007, for information on military and civilian personnel costs.

DATSA, we did a rough estimation based on its availability compared with the REW (on the low end) and the other ATS. Columns 3, 5, and 10 show what the per-tester maintenance costs are for these multiples for each legacy ATS. The total per-tester maintenance cost are the costs in Table 4.2 multiplied by the number of each type of legacy ATS in operation.[9] The total per-tester maintenance cost for each ATS type can be reduced by these amounts by eliminating a single ATS.

Baseline per-type sustainment costs were provided by 565 Combat Sustainment Squadron at Warner Robins ALC with considerable effort, for which we are grateful. The cost data covered all the legacy systems we were analyzing except DATSA. Since DATSA has reportedly been having severe sustainment problems, we set its per-type sustainment cost equal to that for the REW, the most costly of the other five legacy systems. Table 4.3 gives the baseline annual per-type sustainment costs. The other columns give the annual per-type sustainment costs for multiples of the baseline; these are the multiples we will use in the description of our analysis.

Rehosting Costs

Rehosting a set of UUTs on any NATS entails two major costs. First is the cost of the hardware. The quantity to be bought depends on the transferred workload (total work hours) and the availability and capability of the new tester. Typically, new testers are more reliable than older testers, sometimes dramatically so if the legacy ATS is more than a decade old. New testers are also usually technically more capable, with faster and much more powerful computers and instruments. However, because the UUTs are usually of the same technical vintage as the ATS being replaced, that extra capability may not translate into faster testing. To be conservative, we assumed that the test-and-repair times for all UUTs of our notional new tester were the same for the legacy systems. We then assumed a $1.5 million purchase cost for one NATS tester, which is in line with the cost of VDATS and ADTS.[10] We further assumed that the B-1B maintenance communities currently have all the legacy testers they need for the current workload, so there is no need to purchase additional ones.

The second and more significant cost is to transfer the TPS to the NATS and to build hardware interfaces. ITAs in general are relatively inexpensive, on the order of $40,000.[11] However, a number of issues can cause this cost to vary. Many legacy ITAs were "active," i.e.,

Table 4.3
Per-Type Sustainment Costs for Legacy ATS ($M)

ATS	Baseline	2	5	10
DATSA	2.130	4.260	10.650	21.300
DIG	1.138	2.276	5.690	11.380
REW	2.130	4.260	10.650	21.300
DAV	1.138	2.276	5.690	11.380
EPCAT	0.513	1.026	2.565	5.130
USTB	0.080	0.160	0.400	0.800
Range	0.08–2.1	0.16–4.3	0.40–10.7	0.80–21.3

---

[9] MILP computes the number of each of the six types of legacy ATS needed to service the B-1 workload.

[10] Oklahoma City ALC visit, March 26, 2009; Warner Robins ALC visit, May 18, 2009.

[11] Discussion with personnel from 581 SMXS at Warner Robins ALC, May 18, 2009.

had functioning electronics, such as actual LRUs or SRUs (which themselves required maintenance). Current practice is to make new ITAs passive, serving only to route signals from the ATE to the UUT; this requires building a new ITA and the attendant engineering costs. When the need for rehosting has been urgent, legacy ITAs have been adapted with a simple interface to the new ATE (so-called "thin mint" ITAs). Both of the last two types of ITA are substantially cheaper to make than active ITAs.

TPS software rehosting costs have been much larger, ranging up to $2.6 million for LRUs and in the hundreds of thousands for many SRUs.[12] A number of factors drive these costs. First, some UUTs lack good documentation, so that rehosting may require substantial reverse engineering to understand exactly what the UUT does and how it needs to be tested. Second, it is desirable to rewrite legacy TPSs written in older languages, such as ATLAS, in a more-modern language. Third, a legacy TPS may have limited utility because the capabilities of its modern equivalent are very different, thus requiring completely new programs.

As with other costs, the data for estimating these for candidate UUTs are incomplete and fragmentary, for several reasons:

- Relatively few UUT rehostings to the latest ATS generation are complete, so experience with TPS development (and ITA construction) is correspondingly limited.[13]
- The ALCs undertook some of these rehostings themselves, making it hard to separate the costs out from the rest of the ALC budgets.
- In some cases, contractors undertook the work, bundling more than one item together as a single work package at a single price regardless of varying complexity. ADTS at Oklahoma City ALC is one example. In these instances, it is difficult to determine the costs of individual translations.

These factors, together with the uncertainty about documentation we mentioned earlier and the known accuracy limitations of existing TPS cost-estimating models, make projecting rehost costs somewhat problematic.

We assumed a cost of $1 million for LRUs and $300,000 for SRUs,[14] including any preparatory analysis, TPS rewriting, and ITA construction. We did a sensitivity analysis on the translation costs using costs of 0.5 baseline ($500,000 LRU, $150,000 SRU) and 0.1 baseline ($100,000 LRU, $30,000 SRU).

For per-tester maintenance and per-type sustainment costs, we assumed that the NATS costs were zero. This is a conservative assumption that favors rehosting and casts the legacy ATS costs as the excess over similar costs for the NATS.

---

[12] Discussion with ADTS project manager at Oklahoma City ALC, March 26, 2009.

[13] As of March 2009, there were approximately 18 TPSs completed for ADTS. ADTS has a substantially larger number of TPSs in the rehosting pipeline (discussion with ADTS project staff, Oklahoma City ALC, March 26, 2009). The numbers for VDATS were roughly similar at the time of writing (discussion with Warner Robins ALC personnel, May 18, 2009). In comments on a previous draft, personnel in Warner Robins ALC/GRN noted that there is now much more experience with TPS translation (about 156 translations completed by March 2011). This experience could provide a more detailed picture of translation costs that can be incorporated into future roadmaps using the methodology described in this report.

[14] We checked this baseline cost with personnel at Oklahoma City ALC and Warner Robins ALC, and they agreed that it was a reasonable starting point.

Additional Assumptions

To fully specify the parameters for the mixed integer program we used to compute the rehosting decisions, we further assumed the following:

- The new tester can (potentially) test any UUT on the legacy testers.
- The period of evaluation is ten years, with a discount rate of 2.4 percent.[15]
- The NATS purchases and translations are done in the first year and paid in the second; the other legacy ATS costs are discounted across all ten years.
- There is no budget constraint on the rehosting costs.
- The legacy testers are currently able to completely satisfy the maintenance demand for the associated UUTs. We therefore computed the number of legacy testers needed to handle the annual workload for our selected set of UUTs (more may be in service to support other workloads) and assumed these were available. The initial legacy ATS set consists of five DATSAs, three DIGs, 13 REWs, 16 DAVs, two EPCATs, and seven USTBs.[16]
- With our specified flying-hour program for the B-1B fleet as the baseline, the total annual workload is 129,000 hours.

Exploring the Roadmap Decision Space

For our example, the decision we will explore is which UUTs to rehost to NATS. We expect that, as annual per-tester maintenance and per-type sustainment costs increase, more UUTs will be selected for rehosting. The question is how high the legacy ATS costs have to be to drive significant rehosting. If a significant percentage of UUTs are rehosted when the per-tester and per-type support costs are small or reasonably moderate, rehosting to NATS will be economically justifiable. On the other hand, if the support costs have to be quite high before the MILP indicates that many UUTs should be rehosted, we would conclude that rehosting is not economically justified.

The discussion uses two primary metrics to quantify the rehosting. The fraction of work transferred to NATS refers to the part of the total workload of 470 UUTs that is rehosted at a given set of cost values. The second metric is the total individual UUTs rehosted, which can illuminate the determinants of the rehosting decision. In addition to these, we also look at the workload shifted from each legacy ATS to get insights into the structure of the problem.

Rehosting to NATS as a Function of Support Costs

Figure 4.3 illustrates the rehosting decisions computed by the MILP as a function of both per-type and per-tester support costs. Each point in the graph represents a run of the MILP at a particular combination of these costs (the points are joined by lines to help visualize the trends). The x-axis has the per-type costs increasing from baseline to 10 times baseline for each ATS type (each different type will have a different per-type cost, of course). Each colored

---

[15] The ten-year real interest rate is from OMB, 2008.

[16] Note that this is an assumption, given limitations and inconsistencies of data sources. Air Force reviewers indicated that the DIG and REW workload is now being partially done by contractor support because of unreliability issues and that there are only five USTBs currently in use.

Figure 4.3
Rehosting to NATS as a Function of Per-Type and Per-Tester Costs

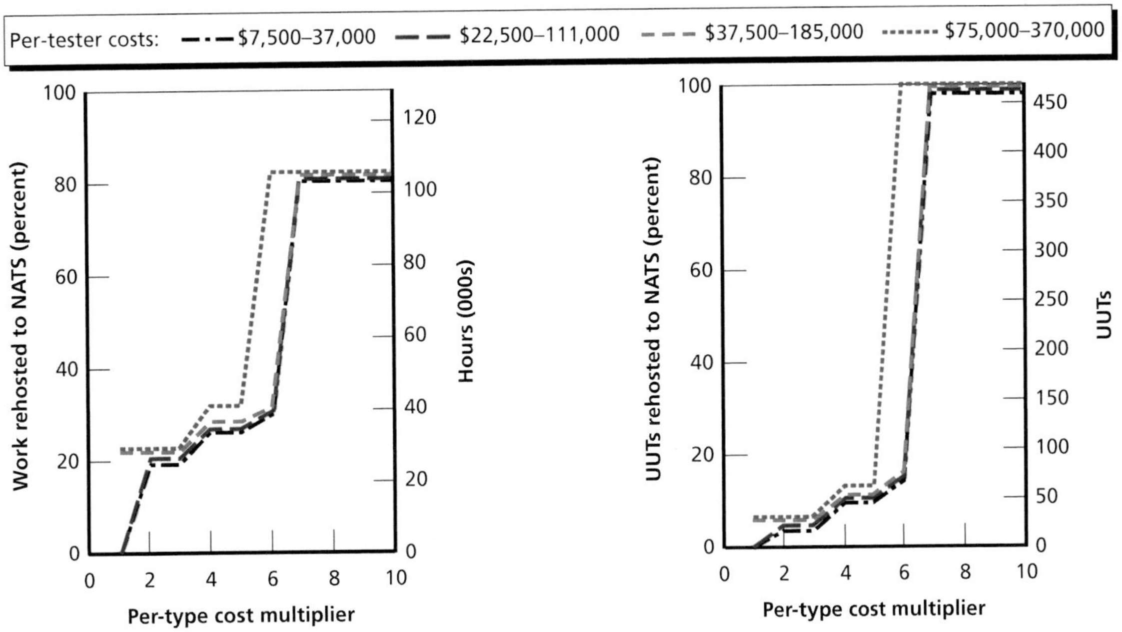

RAND *TR1147-4.3*

line has a different set of per-tester maintenance costs, beginning with the baseline ($7,500–$37,000) and ranging up to $75,000–$370,000 (in multiples of 3, 5, and 10).

At these baseline per-tester and per-type costs, it is not optimal to rehost any UUTs on NATS. However, at double the per-type costs, it is optimal to move 20 percent of the UUT workload to NATS (about 20 UUTs). Note that, even at 10 times the baseline per-type costs, it is optimal to move only 80 percent of the total workload.[17] It is instructive to see which legacy ATS UUTs will be rehosted on MILP.

As Figure 4.3 indicates, no workload would move at the baseline costs. However, doubling the per-type costs or hitting the higher end of the per-tester costs makes it optimal to unload all the UUTs on the REW. The other four older testers follow at per-type multiples ranging from 4 to 7; even DATSA's 350 UUTs are all eventually rehosted (all are SRUs, however, and so their translation costs are cheaper). Only the USTB, the most modern and most reliable legacy ATS with the lowest cost for both per-type maintenance and per-tester sustainment, would not have any work moved from it.

The two lowest model runs in Figures 4.3 and 4.4 were at multiples of 1 and 2 for the annual per-type sustainment costs. We reran the model with fractional multiples from 1 to 2 in steps of 0.1 (1.0, 1.1, etc.) to see exactly where the rehosting started. Figure 4.5 shows the results.[18]

---

[17] On the right-hand graph, the percentage of UUT types migrated is close to 100 percent, but the four UUTs on the USTB are not rehosted to NATS.

[18] The points show the results for an actual run, while the lines are drawn between the points to show the overall trend. Since the graphs in Figure 4.3 were done only at per-type cost multipliers of 1 and 2, the finer structure in Figure 4.5 was not visible.

Figure 4.4
Workload Rehosted from Legacy ATS

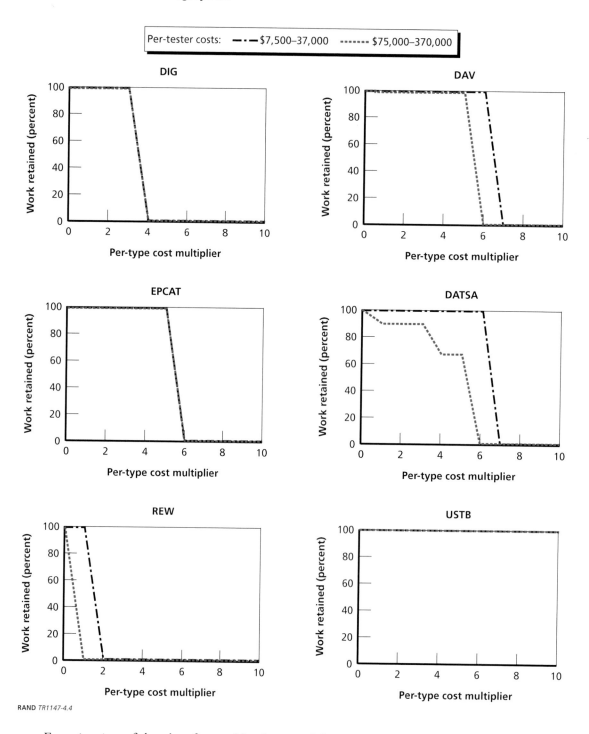

RAND *TR1147-4.4*

Examination of the plots for workloads moved from individual testers (not shown) reveals that some UUTs are moved from DATSA but that the REW workload is completely rehosted when the per-type costs are multiplied by 1.6.

Figure 4.5
Rehosting to NATS as a Function of Per-Type and Per-Tester Costs, with Fractional Multipliers

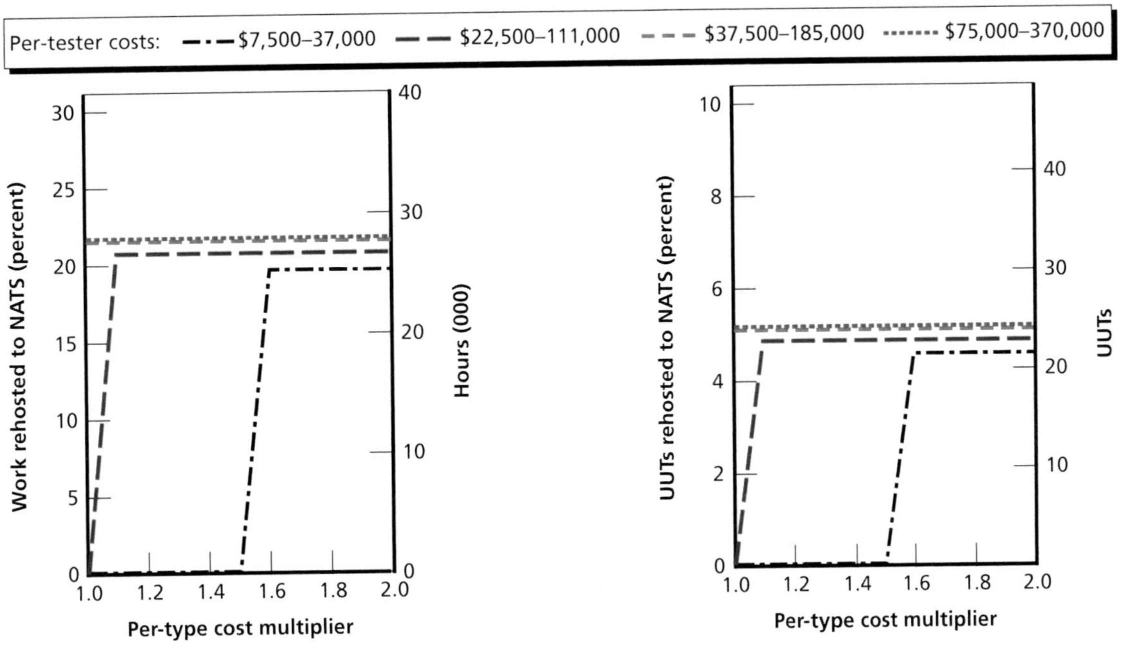

RAND TR1147-4.5

The Effects of Reducing Translation Costs

TPS translation costs are a large component of the cost of rehosting a UUT. It is of interest, therefore, to see how our results would change if translation costs were reduced. Figures 4.6 and 4.7 show the effects of reducing translation costs on workloads and UUTs rehosted, respectively.

Clearly, reducing the translation costs would move the roadmap solution significantly toward rehosting. Only USTB is largely unaffected until its per-tester and per-type support costs get quite high *and* the translation costs are reduced to one-tenth of the baseline.

Increasing the Time Horizon

One of our baseline assumptions was to set the time horizon to ten years. Increasing that adds maintenance and sustainment costs, which favors rehosting, but discounting down weights costs that occur out in the future. For example, with a 2.4-percent discount rate, an annual expenditure of $1 million over 20 years has a net present value of $15.7 million. Figure 4.8 shows how the fraction of work shifted to NATS and the number of UUTs shifted look when the time horizon is extended to 20 years, assuming the baseline translation costs.

Comparing Figure 4.8 with Figure 4.3 shows that, with a time horizon of 20 years, 20 percent of the workload should be rehosted to NATS at baseline costs. This amount rapidly increases to 80 to 100 percent with per-type multipliers of 4 or more.

Figure 4.6
Effects of Translation Costs on Percentage of Workload Rehosted to NATS

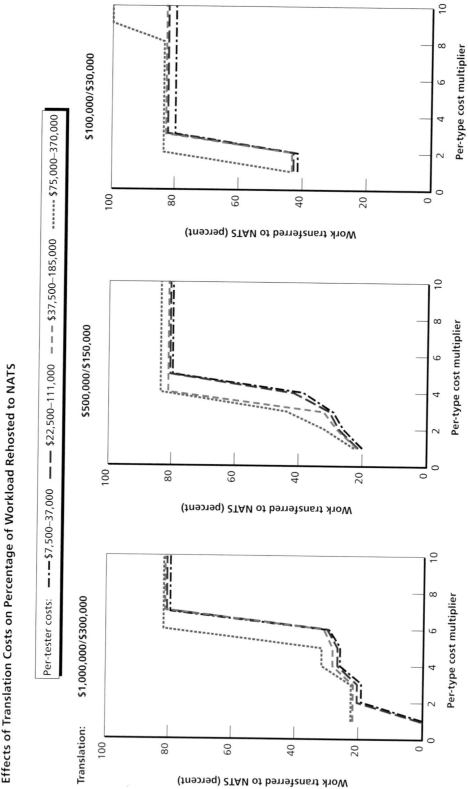

RAND *TR1147-4.6*

Figure 4.7
Effects of Translation Costs on UUTs Rehosted to NATS

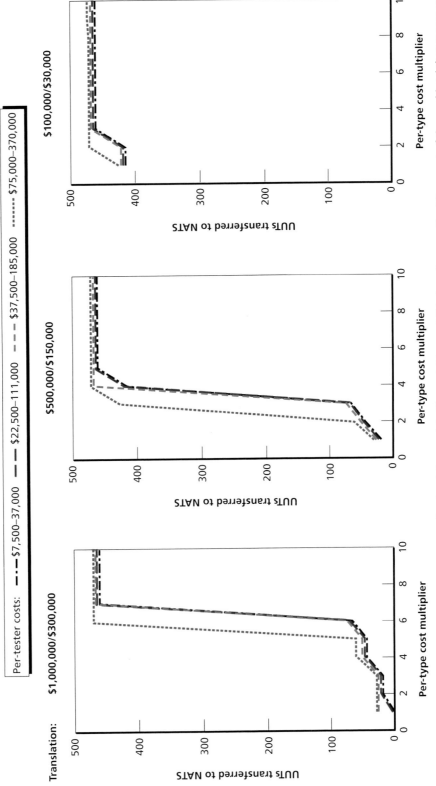

NOTE: The jump in the highest per-tester cost curve for the right graph in Figure 4.6 is not visible here because the 20 percent of the workload that moves at these cost points is the four UUTs from the USTB and this jump is not visible on this scale.

RAND *TR1147-4.7*

Figure 4.8
Effect of 20-Year Horizon on Rehosting to NATS

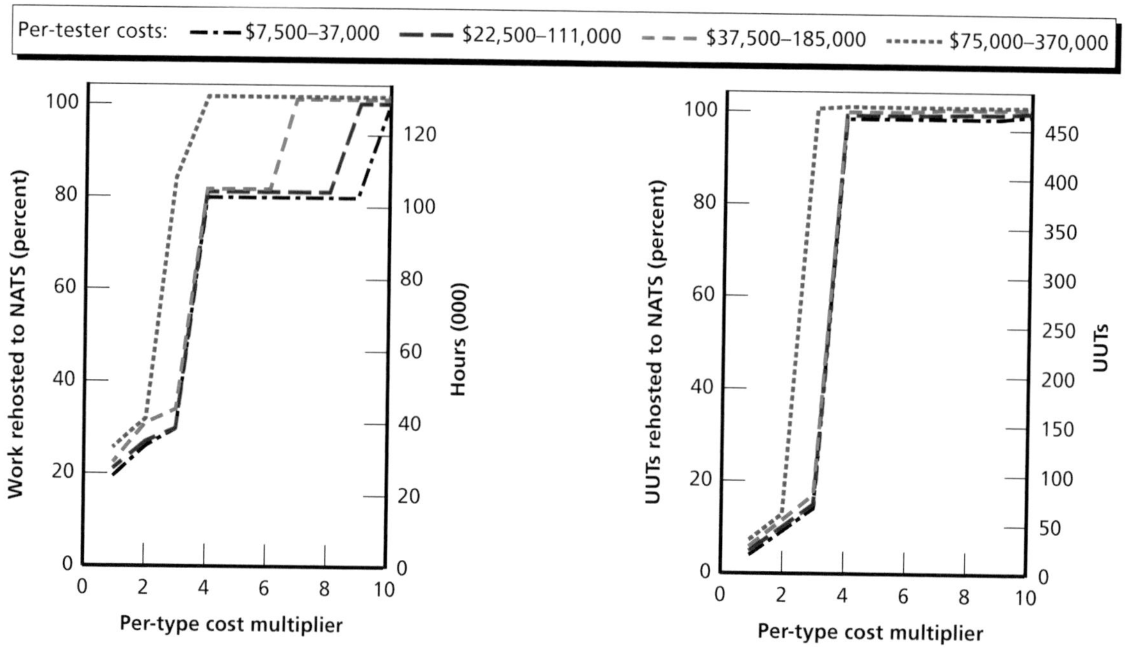

RAND *TR1147-4.8*

Cost-Effectiveness of Rehosting UUTs from a Legacy B-1B ATS

At current baseline estimated per-type and per-tester and translation costs and with a ten-year planning horizon, our analysis of this selected set of avionics UUTs and legacy testers shows that it is not yet cost-effective to rehost workload from five of the six B-1B legacy ATSs. REW is the potential exception. Rehosting the other legacy testers would require increasing the per-type support costs multiples of 4 to 8 for the various types, with the exception of USTB, whose UUTs are not rehosted except at very high per-type and per-tester support costs. At the 20-year horizon, the picture is somewhat different: It becomes cost-effective to rehost about 20 percent of the UUT workload to the NATS, even at baseline costs, and factors of 3 to 6 for per-type costs or 5 to 7 for per-tester costs would make rehosting of 80 percent or more cost-effective.

The case for rehosting the REW workload rests on its baseline per-tester and per-type costs, which are quite close to making the rehosting of its workload cost-effective (MILP computes that rehosting of the REW workload would be cost-effective at 1.6 times the baseline per-type sustainment costs). Given the softness of the baseline per-tester and per-type costs and the possibility that the latter are, in fact, understated, beginning REW rehosting seems prudent.

We noted earlier that the UUT workload distribution on REW is skewed. Figure 4.9 shows the REW workload from Figure 4.2. If only the top three UUTs are rehosted, the overall B-1B workload for this tester will be cut in half. We assumed that per-type sustainment costs are incurred as long as one tester of a given type is operating, but with a substantial

Figure 4.9
REW Workload

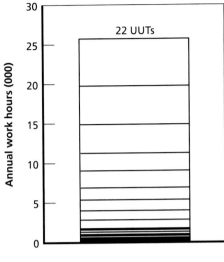

decrease in workload, it is possible that the idle testers could be available for cannibalization to supply parts for active ones and thus help cut sustainment costs. Our ALC interviews indicated a preference to keep all testers fully operational, but for a very old ATS on the verge of retirement, cannibalization might be attractive.[19]

Software translation costs are key drivers of the economic desirability of rehosting but are also relatively uncertain. The baseline translation costs—$1 million for LRUs and $300,000 for SRUs—are consistent with our interviews and the very limited data we have on TPS translations.[20] As noted earlier, software translation is determined by a several factors specific to the UUT: its function, its complexity, and the available documentation. We are cautiously optimistic that more information would allow more-accurate estimates of the rehosting cost for a particular UUT, which would in turn affect the ordering of rehosting decisions. However, we did not have information on the rehosting costs for individual UUTs, except in a handful of cases.

---

[19] This would require storage and warehouse space, which might not be easily available at some facilities.

[20] Refers to interviews at Oklahoma City ALC and Warner Robins ALC (the latter in both the ATS PGM office and SMXG).

CHAPTER FIVE

Conclusions and Recommendations

Building a Roadmap for ATS Modernization

We developed an optimization model that reassigns UUT workloads so as to minimize life-cycle costs of avionics maintenance. This approach can be used to plan the ATS roadmap, allocate rehosting budgets, and cost-justify recapitalization of the ATS inventory.

In designing our methodology, we concluded that the proper level of analysis was the ATS type with its associated UUT workload, rather than the MDS. The modernization roadmap would order the rehosting of the UUTs from a particular ATS type when the analysis demonstrated a cost savings, regardless of which MDS each individual type supported. The MDS would enter into the decision only in setting the period over which the analysis extended; no rehosting would normally be done for an MDS that was being retired toward the beginning of that period. Our analysis focused on the B-1B avionics ATS primarily for efficiency in assembling data, but the process can be applied to any ATS set whose UUTs are being considered for rehosting to one or more new testers.

Rehosting Strategies

Assuming that our baseline costs are correct, it would not be cost-effective to migrate all the workload from any of the ATS types we considered in our case study of selected B-1B avionics. (Note that this is a *subset* of B-1B avionics and that some of the testers, notably DATSA, handle UUTs from other platforms.) In the case study, for the decision to rehost a set of UUTs from a given ATS to be economically feasible, the sustainment cost for the ATS type would have had to increase substantially over the current amount, and/or TPS translation costs would have had to decrease substantially.

Thus, our analysis shows that a single ATS type with the following characteristics would make a good candidate for complete rehosting:

- an old test station with high maintenance costs
- growing obsolescence and/or sustainment engineering issues
- a narrow range of UUTs (low software translation costs).

Incremental Transition

In some cases, incremental workload reduction is cost-effective. For REW, rehosting the top three UUTs in terms of workload would reduce the total REW workload by about 50 percent (of our selected UUTs). Our analysis assumed that the per-type sustainment costs for an ATS type would be incurred as long as one unit of a given type was operational, which is consistent with practices at the ALCs. However, when rehosting a small number of UUTs substantially reduces the annual workload on an ATS type, it may be feasible to continue repair of the other, lower-demand UUTs without the same level of per-type sustainment costs using the newly inactive testers for spares.

Opportunistic Transition

From time to time, one or more TPSs need significant software modifications or revisions because of weapon system LRU or SRU modification or legacy ATS sustainment engineering (e.g., replacing older test equipment with new versions). In these cases, software translation costs are essentially sunk, since they must be paid just to keep the UUTs on the legacy ATS. It may be more cost-effective to rehost the UUTs to a NATS and pay that translation cost instead. Over time, this opportunistic strategy may move a significant amount of the test workload to the new equipment, with the advantages noted under the incremental transition strategy.

Costs of Rehosting UUTs from Legacy ATSs

It is important to keep in mind the substantial uncertainty about the support costs for legacy testers, despite the concern about these costs being a primary driver for moving workloads to modernized ATSs (whether common or not).

Software translation costs are key drivers of the economic desirability of rehosting, and they, too, are uncertain. Our assumed baseline translation costs—$1 million for LRUs and $300,000 for SRUs—are consistent with our interviews and with the very limited data we had on TPS translations for B-1B avionics. The determining factors of software translation costs are specific to the UUT—its function, complexity, and documentation, as well as the documentation for the TPS. We are cautiously optimistic that more-detailed information would improve the accuracy of rehosting cost estimates for individual UUTs, which could affect the cost-benefit analysis. Our methodology is flexible enough to use different translation costs for different UUTs. It has also been argued that, as TPS translation proceeds and more experience is gained, translation costs will decline. This is plausible, but there is as yet limited empirical evidence to support this argument because few rehostings have taken place so far.

Another option is automated translation. Preliminary estimates of 80 percent translation (code lines, not cost) were hypothesized during original planning for the transition to such NATSs as VDATS and ADTS. Experience to date has been disappointing.[1] While reducing translation costs by one-half would make rehosting 20 percent of the workload cost-effective, current thinking is that this estimate is fairly optimistic.

Our analysis has not taken into account several other factors that could favor rehosting. As we noted in the introductory chapter, the Navy was able to reduce both manpower and the number of different specialist types it needed when it rehosted its avionics workload to CASS. The Air Force could well have the same experience at the ALCs and at the unit level. The flexibility and redundancy provided by common, modernized ATSs should be valuable, but these

---

[1] Interviews with Warner Robins ALC personnel on May 5, 2009, and with Oklahoma City ALC personnel on March 26, 2009.

qualities are difficult to quantify, would require detailed analysis of depot operations, and would almost certainly be realized past the ten-year horizon. For older platforms, projecting savings from this source of efficiencies is probably not warranted over their remaining lifetimes.

Future ATS Management

The long-term benefits of modernized and common ATSs present a strong case for making common ATS families the foundation of ATS acquisition for future MDSs and for those just coming into the force. Our methodology can calculate a roadmap for rehosting the workload for current MDSs that will be phased out or reduced in the near term (10 to 20 years), but applying the methodology requires paying close attention to the current and projected status of legacy testers.

Catastrophic Failure of Legacy Testers

Our economic analysis did not address the potential for catastrophic failure of a legacy ATS type, a situation in which none of the ATSs remains operable, making it impossible to test and repair any of the associated UUTs. This could occur because of steadily decreasing ATS fleet sizes and/or the inability to acquire or replace some set of key components that all remaining members of the fleet need.[2] In that case, rehosting would be unavoidable, whatever the cost. Further, if the number of UUTs is too large, timely rehosting might be unobtainable at any price because of the existing demands on programmers and other rehosting resources. This eventuality requires management attention on ATS performance and downtime and input from the engineering and acquisition communities. However, interviews indicated that, in the opinion of maintenance managers, catastrophic failure is highly unlikely. Virtually all legacy testers are still usable, although expensive to maintain and with costs possibly increasing as the equipment ages.

Data Issues

Getting even basic data on Air Force testers was complex and difficult. In the end, we had to substitute assumptions and sensitivity analysis for accurate data on several important parameters, such as ATS performance, reliability, and workloads. In supporting one or more common testers, it is essential to understand how ATS problems affected different MDS fleets, know what UUTs are tested and repaired on different ATSs, and what fraction each UUT represents of the workload of each ATS type. Currently, much of this information is dispersed among ALC staffs and uniformed maintainers and is maintained in an ad hoc fashion. Some data are not available at all.

The WATS tool the project assembled (see Chapter Three and Appendix C) is an example of what could be done with existing Air Force data systems and satisfies some of the data requirements, but even that system cannot currently provide information on such issues as the engineering and acquisition activities required to support older test equipment. For central management of ATS to succeed, the Air Force must improve data availability and accessibility for the data we used in our analysis.

---

[2] During one depot visit, we saw an example of this extreme case for a 1960s vintage tester that had been the only tester to repair a particular B-52 part. The company that originally made the tester was astonished that it was still in service and asked to get the carcass for its corporate display area.

MILP Formulation of the ATS Rehosting Problem

Decision Variables

All decision variables are nonnegative. Binary variables are either 1 or 0; integer variables are constrained to be integers.

X_{ij} = Binary: $X_{ij} = 1$ if UUT_i is assigned on ATE_j; else $X_{ij} = 0$

D_{ijt} = Binary: $D_{ijt} = 1$ if cumulative assignment of UUT_i on $ATE_j > 0$ through period t

R_{ijt} = Binary: first assignment of UUT_i to ATE_j is during period t

U_{jt} = Binary: Use of any ATE_j during period t: $U_{jt} = 1$ if $Z_{jt} > 0$; else $U_{jt} = 0$

Y_{ijt} = Hours of repair time of UUT_i assigned to ATE_j during period t

B_{ijt} = Cumulative repair hours of UUT_i assigned to ATE_j through period t

Z_{jt} = Integer: Number of units ATE_j operated during period t

V_{jt} = Integer: Number of new units of ATE_j added in period t

W_{jt} = Total hours assigned to ATE_j during period t.

Data Variables

G_{it} = Total UUT_i repair workload (RepGens) during period t

F_i = UUT_i repair and/or tester time (mean time to repair), in hours

S_{ij} = Cost to translate UUT_i TPS software on ATE_j

C_j = Cost to acquire one ATE_j

K_t = Maximum allowable investment in period t $(K_1 = 0)$

A_{ij} = Binary: A_{ij} = 1 if UUT_i can be assigned to ATE_j; else A_{ij} = 0

O_{jt} = Cost to operate and maintain any nonzero amount of ATE_j in period t

M_{jt} = Cost to operate and maintain one ATE_j in period t

H_{jt} = Cost to operate one hour of test time on ATE_j in period t

P_{jt} = Available hours per ATE_j station in period t

∞ = A very large number

E_j = ATS_j footprint, in square feet

a = Maximum square footage available in the facility

b = Cost per period per square foot of facility space

L_j = minimum number of ATE_j in the facility.

Objective Function

As presented here, the objective function includes the facility costs for housing an ATS set. As noted in the body of the report, our analysis did not include these costs, although the model can also account for these, given appropriate data.

Minimize

$$\sum_{i=1}^{I}\sum_{j=1}^{J}\sum_{t=1}^{T}Y_{ijt}H_{jt} + \sum_{j=1}^{J}\sum_{t=1}^{T}Z_{jt}M_{jt} + \sum_{j=1}^{J}\sum_{t=1}^{T}U_{jt}O_{jt}$$

$$+ \sum_{i=1}^{I}\sum_{j=1}^{J}X_{ij}S_{ij} + \sum_{j=1}^{J}\sum_{t=1}^{T}C_jV_{jt} + b\sum_{j=1}^{J}\sum_{t=1}^{T}E_jZ_{jt}$$

subject to the following constraints:

Constraint (A.1) ensures that all UUT workload must be assigned in each time period:

$$\sum_{j=1}^{J}Y_{ijt} \geq G_{it}F_i \quad \forall i \in I, t \in T \tag{A.1}$$

Constraints (A.2) and (A.3) ensure that in the initial period, $t = 1$, the UUT workload is assigned to legacy ATEs:

$$Y_{ijt} = G_{it}F_i : S_{ij=0} \qquad \forall i \in I, j \in J, t = 1 \qquad \text{(A.2)}$$

$$Y_{ijt} = 0 : S_{ij} > 0 \qquad \forall i \in I, j \in J, t = 1 \qquad \text{(A.3)}$$

Constraint (A.4) fixes the initial period's cumulative hours for each UUT on each ATS, while constraint (A.5) tracks the cumulative hours assigned to the ATEs over each period:

$$B_{ijt} = Y_{ijt} \qquad \forall i \in I, j \in J, t = 1 \qquad \text{(A.4)}$$

$$B_{ijt} = B_{ijt-1} + Y_{ijt} \qquad \forall i \in I, j \in J, t = 2,\ldots,T \qquad \text{(A.5)}$$

Constraint (A.6) makes sure that UUTs are properly assigned to a corresponding workload:

$$\infty D_{ijt} \geq B_{ijt} \qquad \forall i \in I, j \in J, t \in T \qquad \text{(A.6)}$$

Constraint (A.7) and (A.8) track the period of the initial assignment of UUT to ATS:

$$R_{ijt} \geq D_{ijt} - D_{ijt-1} \qquad \forall i \in I, j \in J, t = 2,\ldots,T \qquad \text{(A.7)}$$

$$R_{ijt} \geq D_{ijt} \qquad \forall i \in I, j \in J, t = 1 \qquad \text{(A.8)}$$

Constraint (A.9) tracks the cumulative UUT workload on each ATS; constraint (A.10) guarantees that there are enough ATS hours to perform the workload:

$$W_{jt} \geq \sum_{i=1}^{I} Y_{ijt} \qquad \forall i \in I, t \in T \qquad \text{(A.9)}$$

$$P_{jt}Z_{jt} \geq W_{jt} \qquad \forall j \in J, t \in T \qquad \text{(A.10)}$$

Constraint (A.11) determines the nonzero number of test systems operated in any period:

$$\infty U_{jt} \geq Z_{jt} \qquad \forall j \in J, t \in T \qquad \text{(A.11)}$$

Constraint (A.12) limits the assignments of UUTs to specific ATSs; constraint (A.13) forces workload to be assigned to ATSs that are available for use in that period:

$$A_{ij} \geq X_{ij} \qquad \forall i \in I, j \in J \qquad \text{(A.12)}$$

$$\infty X_{ij} \geq \sum_{t=1}^{T} Y_{ijt} \qquad \forall i \in I, j \in J \qquad \text{(A.13)}$$

Constraints (A.14) and (A.15) ensure that the cumulative number of ATEs of type j purchased does not exceed the number of ATEs operated:

$$V_{jt} \geq Z_{jt} \qquad \forall j \in J, t = 1 \tag{A.14}$$

$$V_{jt} \geq Z_{jt} - Z_{jt-1} \qquad \forall j \in J, t = 2,\ldots,T \tag{A.15}$$

Constraint (A.16) limits the total investment of purchasing ATEs and translation costs in each period t:

$$K_t \geq \sum_{j=1}^{J} C_j V_{jt} + \sum_{i=1}^{I} \sum_{j=1}^{J} R_{ijt} S_{ij} \qquad \forall t \in T \tag{A.16}$$

Constraint (A.17) ensures that the total footprint of the ATEs is less than the total available space of the facility:

$$a \geq \sum_{j=1}^{J} E_j Z_{jt} \qquad \forall t \in T \tag{A.17}$$

Finally, constraint (A.18) enforces a minimum number of ATEs of each type:

$$Z_{jt} \geq L_j \qquad \forall j \in J, t \in T \tag{A.18}$$

UUTs for B-1B Roadmap Case Study

This table identifies the individual UUTs we included in our analysis of the B-1 Automatic Test Station repairs. They are identified by NIIN, the ATS currently required to perform repair to the component, the number of repairs demanded per year (DEM), the average number test station run hours required to repair the component, and the cost of the component.

Table B.1
UUT Data

NIIN	ATS	DEM	Repair Hours	Cost ($)
011999869	DATSA	4.8	5.7	15,421
011887187	DATSA	2.8	3.4	7,114
011887188	DATSA	2.8	2.3	6,100
011887189	DATSA	2.2	3.4	5,452
011898238	DATSA	1.6	26.9	265,000
011933127	DATSA	1.1	3.5	2,205
011933130	DATSA	6.4	4.3	1,507
011933135	DATSA	1.0	2.7	852
011942450	DATSA	0.6	2.3	592
011990727	DATSA	2.1	5.6	12,428
011990728	DATSA	8.0	5.5	15,458
012156397	DATSA	2.8	2.2	456
011999868	DATSA	0.6	2.3	7,288
011874368	DATSA	1.6	1.0	2,427
012006591	DATSA	10.1	5.6	18,927
012006592	DATSA	2.9	4.4	1,236
012009154	DATSA	7.2	6.7	9,912
012029170	DATSA	1.6	12.2	11,949
012029176	DATSA	0.6	3.4	6,900
012090128	DATSA	1.0	6.6	3,501
012091603	DATSA	0.5	4.5	3,506
012106729	DATSA	0.5	4.5	3,100
012113924	DATSA	38.0	4.5	4,802
012115505	DATSA	20.4	4.5	2,685
012115515	DATSA	14.6	4.5	1,690
011998570	DATSA	4.6	4.4	1,569

Table B.1—Continued

NIIN	ATS	DEM	Repair Hours	Cost ($)
011830334	DATSA	15.2	4.4	9,212
011814331	DATSA	0.5	4.5	1,967
011814332	DATSA	8.4	4.5	2,976
011814333	DATSA	15.9	4.5	2,049
011814334	DATSA	21.1	4.5	4,285
011814335	DATSA	19.8	4.5	4,415
011814346	DATSA	0.6	2.3	3,676
011814398	DATSA	20.8	4.4	15,263
011820062	DATSA	1.7	4.4	2,386
011826360	DATSA	1.0	8.9	5,631
011826362	DATSA	1.7	6.7	11,745
011826377	DATSA	1.7	2.3	1,501
011887186	DATSA	3.4	3.4	5,452
011829764	DATSA	3.7	2.3	1,286
011874373	DATSA	2.1	1.3	2,451
011830393	DATSA	11.7	4.5	834
011830395	DATSA	1.0	5.6	3,147
011831930	DATSA	0.6	2.3	1,105
011832535	DATSA	2.7	4.5	5,242
011833776	DATSA	1.6	6.7	26,280
011833777	DATSA	4.3	17.6	194,578
011835122	DATSA	1.6	3.4	5,407
011835123	DATSA	0.5	4.5	3,579
011835124	DATSA	1.0	2.3	1,227
011835130	DATSA	1.1	2.3	3,771
012142536	DATSA	1.6	2.6	691
011829470	DATSA	5.8	3.4	2,197
012153728	DATSA	2.2	4.5	1,680
012115517	DATSA	1.4	4.5	1,605
012149989	DATSA	2.7	4.5	3,696
012150005	DATSA	1.5	4.4	6,695
012150006	DATSA	3.2	4.5	7,899
012150009	DATSA	7.3	4.5	4,505
012151812	DATSA	7.5	4.5	2,932
012151816	DATSA	3.2	4.4	2,847
015234343	DATSA	2.1	4.5	1,167
012151818	DATSA	4.3	4.5	2,892
012151819	DATSA	1.1	2.3	2,594
012151820	DATSA	14.4	4.4	2,500
012147423	DATSA	1.7	2.3	1,923
012152824	DATSA	2.1	4.5	1,432
012147421	DATSA	8.0	4.5	2,923

Table B.1—Continued

NIIN	ATS	DEM	Repair Hours	Cost ($)
012153729	DATSA	1.7	2.2	2,433
012153730	DATSA	0.5	4.5	1,144
012153738	DATSA	0.5	13.2	1,890
012154559	DATSA	0.5	4.5	768
012154560	DATSA	0.5	4.4	698
012156389	DATSA	2.2	2.3	955
012156390	DATSA	3.4	2.3	822
012156391	DATSA	1.1	2.3	564
012156392	DATSA	1.1	1.2	317
012156393	DATSA	2.9	2.3	4,440
012156394	DATSA	1.1	4.4	2,963
012152823	DATSA	0.5	5.5	8,098
012144490	DATSA	10.6	4.4	4,900
011814214	DATSA	7.5	3.4	6,513
012142537	DATSA	5.3	4.5	2,002
012142541	DATSA	13.3	4.4	5,893
012142543	DATSA	10.2	4.4	1,584
012142544	DATSA	5.6	4.5	2,502
012144428	DATSA	7.9	6.6	25,459
012144434	DATSA	15.4	4.5	1,294
012144435	DATSA	5.9	6.6	7,059
012144483	DATSA	7.9	4.5	6,825
012144484	DATSA	3.7	4.4	3,871
012144485	DATSA	2.1	4.5	5,197
012147424	DATSA	1.1	2.3	3,055
012144487	DATSA	5.3	4.5	1,713
012135970	DATSA	0.5	4.4	10,509
012144491	DATSA	0.5	4.5	1,428
012147382	DATSA	4.8	13.2	11,687
012147386	DATSA	1.6	3.4	4,012
012147404	DATSA	3.4	4.5	9,271
012147405	DATSA	8.7	4.5	5,660
012147406	DATSA	7.3	4.5	6,696
012147407	DATSA	6.2	4.5	2,248
012147410	DATSA	4.8	4.5	5,183
012147411	DATSA	2.7	4.5	1,227
012147413	DATSA	1.5	4.5	1,650
012147418	DATSA	3.2	4.5	4,829
012144486	DATSA	8.5	4.4	1,346
011649119	DATSA	16.2	4.5	894
011661326	DATSA	6.7	4.5	2,512
011638466	DATSA	1.1	3.4	1,386

Table B.1—Continued

NIIN	ATS	DEM	Repair Hours	Cost ($)
011638469	DATSA	0.5	4.4	1,372
011640487	DATSA	0.5	4.5	1,494
011640493	DATSA	0.5	4.5	1,568
011640495	DATSA	1.9	4.5	1,980
011640496	DATSA	1.0	4.5	1,222
011641423	DATSA	0.5	4.4	1,474
011641425	DATSA	0.5	3.4	3,306
011641427	DATSA	4.8	4.5	1,771
011641428	DATSA	1.4	4.4	1,709
011638462	DATSA	0.5	2.3	1,444
011649118	DATSA	0.5	2.3	1,345
011623350	DATSA	0.6	4.5	1,067
011649120	DATSA	1.7	3.3	1,275
011649121	DATSA	14.6	4.5	1,849
011650346	DATSA	0.5	4.4	2,128
011650347	DATSA	1.1	4.4	2,420
011650348	DATSA	1.1	4.5	2,005
011661314	DATSA	7.6	10.2	3,200
011661315	DATSA	33.2	4.5	2,789
011661319	DATSA	6.2	4.5	1,448
011661320	DATSA	19.7	4.5	1,762
011661321	DATSA	0.5	3.4	1,625
011814330	DATSA	1.1	3.4	1,714
011644896	DATSA	1.0	3.4	2,373
011571639	DATSA	1.1	2.3	994
011425604	DATSA	9.6	4.5	32,114
011566072	DATSA	1.7	2.3	1,274
011566076	DATSA	0.5	2.3	1,101
011566720	DATSA	0.6	4.5	1,024
011566727	DATSA	0.5	2.3	1,183
011566728	DATSA	0.6	2.3	815
011566729	DATSA	7.3	2.3	1,927
011569708	DATSA	0.5	2.3	1,322
011570222	DATSA	1.0	4.5	2,314
011570223	DATSA	1.0	2.3	1,779
011570228	DATSA	0.6	4.5	1,921
011638464	DATSA	1.0	2.3	1,586
011570232	DATSA	0.5	2.3	1,638
011661327	DATSA	2.8	4.4	1,832
011599686	DATSA	0.5	13.2	8,068
011601700	DATSA	2.1	5.5	7,204
011601701	DATSA	0.5	4.4	7,560

Table B.1—Continued

NIIN	ATS	DEM	Repair Hours	Cost ($)
011601702	DATSA	0.5	8.9	13,119
011603432	DATSA	3.7	18.6	384,699
011606560	DATSA	12.3	8.9	3,266
011606562	DATSA	1.6	8.9	8,962
011607266	DATSA	0.5	12.8	12,248
011607268	DATSA	0.6	12.8	11,672
011609427	DATSA	6.5	5.6	5,675
011614918	DATSA	1.7	5.6	26,410
011570231	DATSA	1.0	4.5	2,332
011803139	DATSA	6.9	4.5	1,559
011661325	DATSA	27.0	4.5	3,099
011794007	DATSA	8.4	6.7	2,472
011796987	DATSA	8.5	4.5	3,868
011796990	DATSA	1.6	4.4	2,197
011796994	DATSA	1.6	4.4	782
011798443	DATSA	1.1	4.5	4,510
011798478	DATSA	23.7	4.5	2,068
011799684	DATSA	13.5	5.6	3,154
011799685	DATSA	7.5	4.5	602
011799686	DATSA	7.5	2.2	1,386
011802140	DATSA	3.2	8.9	25,625
011793982	DATSA	3.7	4.5	2,221
011802184	DATSA	20.8	8.9	6,941
011793981	DATSA	2.7	4.4	1,677
011803140	DATSA	2.1	4.5	3,141
011803141	DATSA	1.0	2.3	1,549
011806111	DATSA	7.5	2.3	1,815
011806298	DATSA	9.1	4.4	933
011806305	DATSA	3.7	4.5	1,309
011806306	DATSA	1.0	4.5	2,721
011806358	DATSA	5.3	19.3	54,836
011807465	DATSA	4.8	4.4	1,264
011807496	DATSA	8.0	4.5	8,256
011807558	DATSA	11.2	4.5	2,655
012151817	DATSA	6.4	4.5	2,827
011802142	DATSA	1.0	4.5	1,116
011787810	DATSA	4.8	4.5	879
011661328	DATSA	0.5	4.4	1,320
011661330	DATSA	31.0	4.5	3,415
011661331	DATSA	25.9	4.5	1,926
011661335	DATSA	1.1	4.5	1,752
011664225	DATSA	0.5	2.5	1,542

Table B.1—Continued

NIIN	ATS	DEM	Repair Hours	Cost ($)
011670368	DATSA	1.1	2.3	1,342
011674507	DATSA	5.6	6.6	2,899
011674508	DATSA	3.4	6.7	3,053
011674509	DATSA	2.8	4.5	4,632
011674510	DATSA	1.7	4.5	2,751
011681366	DATSA	3.6	4.4	1,745
011794006	DATSA	0.5	4.5	1,595
011787809	DATSA	11.7	4.5	1,574
011814215	DATSA	3.2	3.4	2,365
011791781	DATSA	17.0	4.5	3,330
011791782	DATSA	19.1	4.5	4,295
011791783	DATSA	18.9	3.4	3,203
011791784	DATSA	7.2	4.5	1,485
011791787	DATSA	1.1	4.2	785
011791847	DATSA	2.2	2.2	1,188
011791848	DATSA	6.7	3.4	1,028
011793638	DATSA	6.9	4.5	1,045
011793976	DATSA	4.5	4.4	1,006
011793977	DATSA	14.0	13.2	7,808
011793978	DATSA	0.5	4.5	1,072
011780542	DATSA	2.1	5.2	876
012649895	DATSA	1.1	3.4	1,136
012621456	DATSA	0.5	2.3	1,644
012628375	DATSA	0.5	2.7	2,560
013216836	DATSA	66.9	4.5	7,100
013087080	DATSA	1.1	4.5	1,284
012632484	DATSA	15.5	4.1	66,474
012639954	DATSA	7.9	4.5	3,855
013744885	DATSA	1.0	4.4	2,145
013080909	DATSA	23.0	4.5	8,902
013243584	DATSA	0.5	4.4	1,820
013080908	DATSA	38.8	4.5	8,939
012650980	DATSA	0.9	2.1	13,445
013075243	DATSA	1.6	6.7	1,552
013072615	DATSA	47.8	4.5	10,000
012663505	DATSA	1.1	2.3	1,299
013061299	DATSA	31.5	2.3	6,716
012639957	DATSA	37.8	4.5	10,376
013594722	DATSA	3.2	4.4	38,015
012572850	DATSA	1.1	4.4	1,355
012790882	DATSA	9.6	4.0	16,174
013658267	DATSA	1.1	4.5	1,340

Table B.1—Continued

NIIN	ATS	DEM	Repair Hours	Cost ($)
012920815	DATSA	1.0	1.7	3,705
012575484	DATSA	4.7	24.0	9,824
012579452	DATSA	3.2	11.0	10,591
012621455	DATSA	1.0	2.3	1,510
013619954	DATSA	8.4	4.5	7,421
012620507	DATSA	3.7	6.7	4,173
013591729	DATSA	0.5	4.5	3,890
012591584	DATSA	3.7	29.8	6,830
013524130	DATSA	0.5	4.5	1,768
013476724	DATSA	20.2	4.5	18,249
012620506	DATSA	1.1	4.5	7,907
013258093	DATSA	6.9	4.4	6,593
012693640	DATSA	1.0	2.3	1,577
013658264	DATSA	0.5	4.5	7,745
012908622	DATSA	3.4	4.4	4,808
012663508	DATSA	1.1	2.3	2,288
012905886	DATSA	3.9	5.6	5,904
012908614	DATSA	1.1	4.5	13,726
012925795	DATSA	0.5	1.8	3,692
012908615	DATSA	7.3	4.4	2,851
012908620	DATSA	1.7	4.4	3,131
012905883	DATSA	12.4	6.7	18,555
012925524	DATSA	2.8	1.0	1,913
012847254	DATSA	15.2	4.5	2,650
012908623	DATSA	7.9	4.5	3,318
012908624	DATSA	5.1	2.3	3,474
012908625	DATSA	15.7	4.4	3,194
012908626	DATSA	3.9	4.4	2,953
012908630	DATSA	1.1	4.4	3,810
012908631	DATSA	4.5	4.5	5,352
012920814	DATSA	0.5	2.4	2,677
012908621	DATSA	9.0	15.6	3,346
012725994	DATSA	1.6	4.0	20,871
013667287	DATSA	1.1	4.5	2,516
013039377	DATSA	66.1	4.5	3,822
013036927	DATSA	0.5	8.9	4,983
012702933	DATSA	2.1	12.2	13,264
013036926	DATSA	0.5	4.4	5,625
012704819	DATSA	1.1	4.4	2,646
012905885	DATSA	7.9	7.7	5,405
013023458	DATSA	153.0	4.5	3,962
012684664	DATSA	2.2	4.5	1,599

Table B.1—Continued

NIIN	ATS	DEM	Repair Hours	Cost ($)
012735199	DATSA	3.9	4.5	11,472
013023457	DATSA	55.7	4.5	5,009
012156889	DATSA	1.7	4.5	1,236
012151813	DATSA	66.1	4.4	3,161
012925797	DATSA	0.5	1.3	1,882
012843582	DATSA	1.0	2.3	1,286
012846337	DATSA	1.1	4.5	12,394
013030485	DATSA	2.8	4.4	3,962
012265417	DATSA	2.9	4.5	18,293
013745627	DATSA	1.6	4.5	1,685
012210173	DATSA	6.2	3.4	1,125
012211062	DATSA	1.7	5.5	16,032
012213945	DATSA	1.1	4.5	1,724
015090594	DATSA	8.0	13.2	95,626
012221961	DATSA	3.2	4.5	1,240
012199403	DATSA	7.3	4.4	7,422
012244261	DATSA	221.7	4.5	12,003
015234341	DATSA	2.1	4.5	1,441
012265418	DATSA	1.4	4.5	18,030
015022473	DATSA	2.7	12.1	22,509
012286002	DATSA	4.4	4.4	1,822
012290578	DATSA	3.4	17.6	11,683
012344321	DATSA	0.1	3.4	5,005
013658265	DATSA	1.7	4.5	1,566
015022471	DATSA	1.6	11.0	30,399
012233768	DATSA	260.4	4.4	7,666
012180305	DATSA	0.5	3.4	3,107
012156890	DATSA	0.5	4.5	1,129
012160773	DATSA	3.0	4.5	2,583
012163215	DATSA	3.7	5.5	1,392
012166968	DATSA	1.6	4.5	4,017
015234342	DATSA	4.8	4.4	5,758
012177995	DATSA	1.1	3.4	3,475
012199404	DATSA	1.0	4.5	6,192
012180259	DATSA	1.7	6.7	2,742
014764212	DATSA	4.8	4.4	6,258
012184287	DATSA	27.6	4.0	2,113
012189290	DATSA	2.2	4.4	1,960
012189295	DATSA	1.7	4.5	2,496
012198296	DATSA	4.5	3.0	8,946
012198297	DATSA	1.7	2.2	8,370
012198298	DATSA	3.9	4.5	13,011

Table B.1—Continued

NIIN	ATS	DEM	Repair Hours	Cost ($)
012198299	DATSA	1.7	3.4	18,781
012177998	DATSA	1.0	2.3	1,003
012400116	DATSA	3.2	12.0	42,536
012441928	DATSA	1.1	4.5	1,306
013751598	DATSA	2.1	4.4	1,958
012437950	DATSA	1.7	3.4	414
012431129	DATSA	2.1	6.7	598
013754470	DATSA	1.6	4.5	1,470
013754471	DATSA	2.2	4.5	1,388
013782988	DATSA	1.0	4.5	1,811
012349183	DATSA	0.5	1.5	1,060
013751597	DATSA	2.1	4.4	1,401
013751596	DATSA	0.5	4.4	1,401
013876109	DATSA	3.2	4.5	1,454
013876115	DATSA	0.5	4.4	1,606
013754469	DATSA	0.5	3.4	1,476
012510259	DATSA	9.6	10.0	49,744
012395682	DATSA	8.4	4.4	4,640
013745628	DATSA	2.2	4.5	1,992
012562525	DATSA	2.7	24.8	14,270
012390495	DATSA	0.6	4.5	940
012390496	DATSA	0.6	4.5	1,104
012390499	DATSA	0.6	4.5	1,569
013931372	DATSA	2.2	4.5	1,741
012394582	DATSA	6.2	4.4	5,664
012395689	DATSA	0.5	2.3	950
014420950	DATSA	0.5	5.2	1,252
014193719	DATSA	1.6	4.4	3,216
013750386	DATSA	1.6	3.4	1,791
012392951	DATSA	0.6	4.5	1,331
013449239	DAV	39.4	24.1	310,583
011486207	DAV	26.7	11.1	15,064
010363198	DAV	59.3	25.1	214,460
011644913	DAV	62.0	8.9	43,837
011507427	DAV	37.4	13.3	201,300
011507428	DAV	71.6	14.8	320,448
013451109	DAV	98.2	23.3	85,171
011756188	DAV	30.4	10.9	48,985
012931237	DAV	12.3	6.7	33,033
011611087	DAV	36.3	8.9	9,730
011670881	DAV	44.9	10.2	43,930
014397664	DAV	47.0	11.0	15,215

Table B.1—Continued

NIIN	ATS	DEM	Repair Hours	Cost ($)
013076363	DAV	100.3	31.7	43,501
011573914	DAV	65.1	13.3	54,115
013076362	DAV	230.9	12.4	116,282
013451110	DAV	25.1	10.9	141,050
011874063	DAV	302.1	11.0	32,269
011898118	DAV	5.9	12.1	38,798
012628319	DAV	91.2	17.0	137,562
012283603	DAV	8.5	5.6	7,740
011853017	DAV	1.1	8.9	13,742
012823674	DAV	16.0	11.6	132,840
012630536	DAV	133.9	35.6	178,675
011799587	DAV	29.8	17.3	37,868
012630425	DAV	93.0	14.6	41,385
012616069	DAV	137.6	8.9	79,375
012594655	DAV	44.8	14.4	155,660
012112086	DAV	13.3	8.9	36,400
012353510	DAV	70.4	5.4	91,259
012546944	DAV	85.0	17.2	60,428
010351092	DAV	54.0	26.2	100,471
011856507	DAV	3.7	30.7	42,255
012828765	DAV	141.1	28.3	137,000
012754675	DAV	78.4	20.9	135,412
011853016	DAV	14.4	8.9	30,890
011829763	DAV	13.4	8.9	24,300
012768318	DAV	81.1	36.3	178,447
011819872	DAV	58.2	6.7	14,079
012719168	DAV	86.1	29.7	93,823
012658497	DAV	18.2	12.7	135,991
012654025	DAV	12.8	11.2	55,541
012652887	DAV	189.7	19.8	77,377
011829328	DAV	12.3	8.9	23,962
011829329	DAV	19.2	8.9	30,784
011802117	DAV	11.2	26.5	198,634
012695437	DAV	135.4	24.1	146,035
011642197	DIG	6.4	12.8	15,263
011642196	DIG	1.6	9.6	24,038
013751527	DIG	48.5	18.6	36,497
012496118	DIG	2.1	8.9	11,430
012456683	DIG	61.5	8.9	11,200
011590288	DIG	53.3	9.9	20,070
014453687	DIG	109.9	22.0	184,000
014672007	DIG	10.1	14.2	69,455

Table B.1—Continued

NIIN	ATS	DEM	Repair Hours	Cost ($)
011569682	DIG	9.1	7.7	21,516
011569683	DIG	5.3	5.0	3,780
011641411	DIG	3.2	6.6	9,396
011441284	DIG	9.1	10.6	40,836
011425603	DIG	11.8	8.9	52,605
011425606	DIG	29.4	9.2	64,312
011433525	DIG	34.7	11.0	22,257
012581147	DIG	17.6	13.9	106,657
012571244	DIG	32.6	11.2	261,392
012695439	DIG	52.4	19.0	103,938
011433527	DIG	21.9	12.3	18,425
011433526	DIG	13.4	8.9	14,917
012722138	DIG	2.1	16.0	15,739
011477221	DIG	49.7	19.8	155,000
012153504	DIG	3.9	8.9	5,570
012575268	DIG	30.4	8.9	11,640
012575267	DIG	4.8	8.9	10,949
012149769	DIG	8.0	5.6	19,000
011873230	DIG	17.6	3.2	31,724
014438904	EPCAT	14.4	8.8	371,385
014438903	EPCAT	18.7	8.0	140,135
012355183	EPCAT	16.0	9.5	211,441
014829086	EPCAT	20.8	5.5	174,048
015006333	EPCAT	13.4	7.5	204,418
013996877	EPCAT	34.7	11.1	481,934
012185008	EPCAT	7.0	14.3	14,248
014831393	EPCAT	7.5	3.0	100,680
012412204	EPCAT	4.3	9.5	115,806
012704772	EPCAT	12.3	10.0	260,825
013389677	EPCAT	8.5	10.0	152,674
013516079	EPCAT	7.5	10.1	112,663
013648441	EPCAT	16.6	24.3	387,558
012507000	EPCAT	21.4	10.5	161,318
013994174	EPCAT	17.6	11.1	147,843
013994167	EPCAT	27.3	11.1	148,270
013994168	EPCAT	16.0	21.0	350,140
013994170	EPCAT	27.8	26.1	431,218
013994173	EPCAT	19.2	11.1	99,619
013994171	EPCAT	8.0	34.0	225,019
013994172	EPCAT	4.3	11.1	238,331
012562544	REW	48.1	46.5	372,833
012400091	REW	79.0	25.9	254,551

Table B.1—Continued

NIIN	ATS	DEM	Repair Hours	Cost ($)
012177376	REW	16.6	7.4	249,743
012587062	REW	14.4	14.8	110,147
012399984	REW	8.0	7.1	212,238
012177425	REW	25.0	10.4	269,752
012581133	REW	7.0	19.1	330,180
012572861	REW	2.1	1.5	289,090
012404224	REW	37.4	29.3	373,632
012166064	REW	19.2	10.7	107,459
012643364	REW	63.6	19.2	250,701
012571341	REW	134.0	27.5	180,346
012572789	REW	179.8	27.2	363,172
012408402	REW	1.7	8.0	436,096
012398983	REW	59.8	27.5	324,161
012404223	REW	7.5	9.1	291,803
012403755	REW	62.4	20.7	140,163
012403754	REW	216.7	27.4	315,104
012387922	REW	12.3	17.8	130,253
014540015	REW	12.8	13.1	142,761
014540011	REW	15.0	17.9	124,000
012403271	REW	0.5	8.0	290,285
013445855	USTB	16.5	40.1	174,573
014335623	USTB	256.5	39.6	435,457
011507528	USTB	106.1	30.9	221,100
011507527	USTB	329.1	30.9	460,000

The Web Automatic Test System

The data-gathering efforts described in this report yielded a number of databases from ALCs, bases, and other Air Force agencies in the form of text reports, Microsoft Access databases, and Excel spreadsheets. Building relationships among these varied data sets required conversion to Access and cross-linking different tables to resolve ambiguities in multiple identifiers (part numbers, NIINs, and other descriptors). To make this massive amount of data available to our researchers in an interactive, user-friendly environment, we built a prototype web query tool, which we called WATS. The tool allows a user to focus his or her inquiries by selecting web controls in a series of tabbed panels (see Figure C.1). The panels allow queries for parts or equipment via either NSN or part numbers or AN-designators or acronyms (equipment) or related computer program identification number or aircraft type and MDS.

Some examples of WATS queries are ATS identification; UUT to ATS assignments; cross-references between NSNs and NIINs, part numbers, equipment designators, acronyms, and computer program identification numbers; and the ATS equipment inventory (quantity and location and sometimes the owner, such as a wing).

At several points in the project, such organizations as the ATS Program Office and ACC personnel responsible for ATS policy pointed out that WATS could provide the centralized information on ATS that is essential for managing it across different major commands and multiple MDSs.[1] This tool and the relational database behind it, when connected to continuously updated databases, could serve as a model for an ATS decision support system for the ATS Program Office, the major commands, and the Air Staff.

---

[1] RAND made an initial data transfer to the Logistics, Installations and Mission Support–Enterprise View system organization, which showed interest in developing a web-access ATS prototype. This system collects data from a variety of other data systems that it tries to synthesize into a comprehensive overview of the health of the enterprise. See Petcoff, 2010.

Figure C.1
The Prototype Web-ATS Query Tool

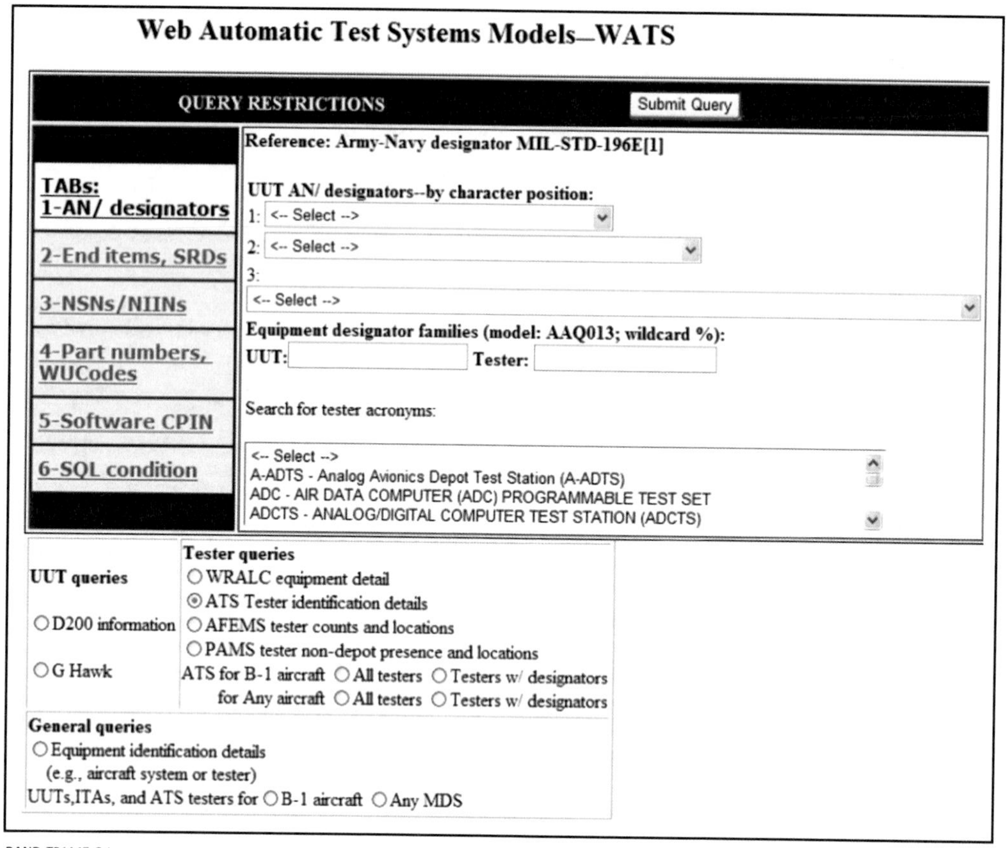

Web Automatic Test Systems Models—WATS

| QUERY RESTRICTIONS | Submit Query |

Reference: Army-Navy designator MIL-STD-196E[1]

TABs:
1-AN/ designators

2-End items, SRDs

3-NSNs/NIINs

4-Part numbers, WUCodes

5-Software CPIN

6-SQL condition

UUT AN/ designators--by character position:
1: `<-- Select -->`
2: `<-- Select -->`
3:
`<-- Select -->`

Equipment designator families (model: AAQ013; wildcard %):
UUT: [   ]   Tester: [   ]

Search for tester acronyms:

`<-- Select -->`
A-ADTS - Analog Avionics Depot Test Station (A-ADTS)
ADC - AIR DATA COMPUTER (ADC) PROGRAMMABLE TEST SET
ADCTS - ANALOG/DIGITAL COMPUTER TEST STATION (ADCTS)

UUT queries

○ D200 information

○ G Hawk

Tester queries
○ WRALC equipment detail
⊙ ATS Tester identification details
○ AFEMS tester counts and locations
○ PAMS tester non-depot presence and locations
ATS for B-1 aircraft ○ All testers ○ Testers w/ designators
    for Any aircraft ○ All testers ○ Testers w/ designators

General queries
○ Equipment identification details
  (e.g., aircraft system or tester)
UUTs,ITAs, and ATS testers for ○ B-1 aircraft ○ Any MDS

RAND *TR1147-C.1*

Bibliography

Air Combat Command, "B-1B ATE ROADMAP," November 2001.

Automated Computer Program Identification Number System (ACPINS), website, undated. As of March 29, 2011:
https://acpins.tinker.af.mil

Automated Test Systems Executive Agent, *DoD Automatic Test Systems Framework Roadmap*, undated. As of September 1, 2011:
http://www.acq.osd.mil/ats/

———, DoD Automated Test Systems Executive Directorate Home Page, September 1, 2011. As of September 1, 2011:
http://www.acq.osd.mil/ats/

Burden, J., P. A. Curry, D. Roby, and F. Love, "Introduction to the Next Generation Automatic Test System (NGATS)," AUTOTESTCON, 2005.

Dahlman, Carl J., *The Cost of a Military Person-Year*, Santa Monica, Calif.: RAND Corporation, MG-598-OSD, 2007. As of March 29, 2011:
http://www.rand.org/pubs/monographs/MG598.html

Defense Logistics Agency, Logistics Information Service, Federal Logistics Data (FEDLOG) Information Center, website, July 1, 2011. As of September 1, 2011:
http://www.dlis.dla.mil/fedlog/

DoD Instruction 5000.2, Operation of the Defense Acquisition System, Washington, D.C.: Office of the Secretary of Defense, 1995.

Eckersley, Joseph, *Business Case Analysis for Automatic Test System Standardization Using the Versatile Depot Automatic Test Station*, Warner Robins Air Logistics Center: Air Force Materiel Command, 2008.

Federation of American Scientists, "B-1 Lancer," web page, undated. As of October 6, 2009:
http://www.fas.org/programs/ssp/man/uswpns/air/bombers/b1b.html

GAMS Development Corporation, the General Algebraic Modeling System (GAMS) home page, 2011. As of December 14, 2011:
http://www.gams.com/

GAO—*See* U.S. General Accounting Office.

Gebman, Jean R., *Challenges and Issues with the Further Aging of U.S. Air Force Aircraft: Policy Options for Effective Life-Cycle Management of Resources*, Santa Monica, Calif.: RAND Corporation, TR-560-AF, 2009. As of March 29, 2011:
http://www.rand.org/pubs/technical_reports/TR560.html

Geller, Amanda B., David George, Robert S. Tripp, Mahyar A. Amouzegar, and Charles Robert Roll, Jr., *Supporting Air and Space Expeditionary Forces: Analysis of Maintenance Forward Support Location Operations*, Santa Monica, Calif.: RAND Corporation, MG-151-AF, 2004.

Higgins, Robert C., *Analysis for Financial Management*, McGraw-Hill, 2007.

Hillier, Frederick S., and Gerald J. Lieberman, *Introduction to Operations Research*, 6th ed., McGraw-Hill, 1995.

Keating, Edward G., and Matthew C. Dixon, *Investigating Optimal Replacement of Aging Air Force Systems*, Santa Monica, Calif.: RAND Corporation, MR-1763-AF, 2003. As of March 29, 2011:
http://www.rand.org/pubs/monograph_reports/MR1763.html

Naval Air Systems Command, "Consolidated Automated Support System," Aircraft and Weapons website, undated. As of September 1, 2011:
http://www.navair.navy.mil/index.cfm?fuseaction=home.display&key=576C1728-0C54-472F-84B3-CA9D40BBF46D

Office of Management and Budget, OMB Circular A-94, Appendix C, 10-Year Real Interest Rate, December 2008. As of December 2009:
http://www.whitehouse.gov/omb/circulars_a094_a94_appx-c

Office of the Secretary of Defense, *Integrated Security Posture (ISP)*, May 2008.

The Open Group, website, 2011. As of September 1, 2011:
http://www3.opengroup.org/

Petcoff, Russell P., "Air Force Program Recognized for Excellence in Government," Air Force News Service, May 5, 2010. As of December 7, 2011:
http://www.af.mil/news/story.asp?id=123203270

Ramey, Timothy L., and Edward G. Keating, *United States Air Force Aircraft Fleet Retention Trends: A Historical Analysis*, Santa Monica, Calif.: RAND Corporation, TR-740-AF, 2009. As of March 29, 2011:
http://www.rand.org/pubs/technical_reports/TR740.html

U.S. Air Force, *USAF Computer Program Identification Numbering (CPIN) System*, user's manual, TO 00-5-17, May 15, 2003a. As of March 29, 2011:
http://www.everyspec.com/USAF/USAF+-+Tech+Manuals/TO_00-5-17_4009/

———, *USAF Automated Computer Program Identification Number System (ACPINS)*, software manager's manual, TO 00-5-16, October 15, 2003b. As of March 29, 2011:
http://www.everyspec.com/USAF/USAF+-+Tech+Manuals/T--O--_00-5-16_4008/

U.S. Department of Defense, *Mandatory Procedures for Major Defense Acquisition Programs (MDAPS) and Major Automated Information System (MAIS) Acquisition Programs*, DoD Instruction 5000.R, April 5, 2002.

U.S. General Accounting Office, *Military Readiness: DOD Needs to Better Manage Automatic Test Equipment Modernization*, Report to the Chairman, Subcommittee on National Security, Emerging Threats and International Relations, Committee on Government Reform, House of Representatives, Washington, D.C., GAO-03-451, March 2003. As of March 29, 2011:
http://www.gao.gov/new.items/d03451.pdf

Under Secretary of Defense for Acquisition and Technology, "DoD Policy for Automatic Test Systems (ATS)," memorandum, April 29, 1994.

Wynne, Michael, Under Secretary of Defense for Acquisition, Technology, and Logistics, memorandum, July 28, 2004.